IMAGES
of America

MELBOURNE'S LOGGING ERA
1912–1932

America's lumber industry was born around 1631 in South Berwick, Maine, and Maine was the center of the industry for over 200 years. The center shifted to New York by 1850, Pennsylvania by 1860, and the Great Lakes states by 1892. Animal-powered transportation depended on snow, so felling began in the late summer or early fall and continued until mid- to late December when the snow became too deep. In late 1876, the steam locomotive revolutionized the logging industry. Lumbermen were no longer dependent on the whims of winter snows or spring floods to ship their timber to the mill. By 1909, the South had become the center of activity. There, felling could be carried on year-round, as weather seldom interfered. (Author's collection.)

On the Cover: At the center of a typical logging "set," somewhere within the vast Jane Green Swamp, a portable overhead skidder is pulling logs to the "landing" beside a temporary spur, where they will remain until loaded onto railway cars. (Author's collection.)

ISBN 978-1-5316-5848-9

Published by Arcadia Publishing
Charleston, South Carolina

Library of Congress Control Number: 2010936249

For all general information, please contact Arcadia Publishing:
Telephone 843-853-2070
Fax 843-853-0044
E-mail sales@arcadiapublishing.com
For customer service and orders:
Toll-Free 1-888-313-2665

Visit us on the Internet at www.arcadiapublishing.com

*"There is nothing new in the world,
except the history you do not know."*

—Harry Truman

IMAGES
of America

MELBOURNE'S LOGGING ERA 1912–1932

Ed Vosatka

ARCADIA PUBLISHING

CONTENTS

ACKNOWLEDGMENTS

Few company records survived, and most of Melbourne's early newspapers burned in 1919. Melbourne, Eau Gallie, and Deer Park news printed in other newspapers helped fill in the gaps. Legal deeds, plats, and surveys in the archives of Brevard and Osceola Counties provided detailed information in those fields. Patents, technical works, and texts on logging methods, railroad construction, and steel structural engineering, published in the early 1900s, provided the means to interpret and understand the technology of those days.

I am indebted to everyone who has helped me over the past 30-something years. They include the following former company employees and residents who had some interaction with the mill and its railway: Ray Bush, Norman Coates, Martha Edwards, Rossie Embry, Myrtice Floyd, Frances Henley, Harry Hubler, Bentley Hughes, laborer Bob Kempfer, teacher Sallie Madden, paymaster Bob Rowe, engineer Ernest "Frog" Smith, teacher Mary Ellen Smith, engineer Vertner Snow, Francis Snow, and teacher Myrtle Svedelius; descendants of the original Hopkins and Kempfer families: Margie Barrett, Grace Blake, Katherine Doran, Helen Fairman, Wendy Mann, Carolyn Kempfer, Billy Kempfer, Reed Kempfer, and their families; and our own local historians and businesspeople: Chuck Baird, Jim Culberson, "Peb" Elliott, Carol Glanton, Johnny Hiott, Fred Hopwood, Charles Jackson Sr., Gordon Patterson, Dennis and Kathy Meehan, Karen Raley, Ken Snyder, and Frank Thomas. Several others who contributed photographs are acknowledged in the text.

There are also librarians and archivists from Brevard and Osceola Counties, Brevard County Historical Commission (BCHC), Brevard County Libraries, City of Melbourne, Florida East Coast Railway (FEC), Florida Game and Freshwater Fish Commission (GFC), Florida Historical Society (FHS), Florida State Archives, Florida Power and Light (FPL), Library of Congress Geography and Map Division, Macedonia Missionary Baptist Church Board of Directors, NOAA/National Climatic Data Center, Railroad Museum of Pennsylvania, Smithsonian Museum of American History, St. Johns River Water Management District (SJRWMD), South Brevard Historical Society (SBHS), and the Southern Museum of Civil War and Locomotive History.

Most of all, special thanks go to my favorite reference librarian, Diane, whom I met while doing this research and who is now my wife and best friend.

INTRODUCTION

Florida's entire peninsula was originally a giant forest, with the exception of the Everglades. Its forest resources caught the US government's attention in the 1790s as a source of live oak for ships in its fledgling navy.

An 1870 federal census study showed the area of Florida was 59,268 square miles, with about 6.3 percent of this being farmland, loosely classified as "improved, improved woodland, or unimproved woodland." Between 1850 and 1870, the value of Florida's lumber products rose from $391,034 to $2,235,780.

A board foot, the standard measure of lumber, is the equivalent of a 12-by-12-by-1-inch board. Thinking in terms of a standard 12-inch library bookshelf, 5,280 board feet would make a shelf one mile long. The Library of Congress has 530 miles of shelving, which, placed end to end, would reach from Melbourne to Birmingham, Alabama.

Florida's entire lumber industry peaked in 1909, manufacturing 1,201,734,000 feet, enough to build a shelf from Melbourne to the surface of the moon. That year, the US Department of Agriculture (USDA) reported 471 active mills in the state, more than in any other year. Less than a handful of species supplied nearly all the lumber cut, being 99.6 percent softwoods (yellow pine, cypress, and cedar) and 0.4 percent hardwoods (red gum, white oak, white ash, yellow poplar, hickory, red oak, and maple). Free on board (FOB) is a shipping term denoting that the price quoted includes loading onto a vehicle at the mill.

Florida's yellow pine industry also peaked in 1909, with the USDA reporting 1,110,840,000 feet manufactured (8.5 times around the world), placing the state ninth in the nation. Florida yellow pine averaged $13.23 per thousand feet, FOB mill, compared to a $12.69 national average. The best price for Florida pine, however, was in 1920, averaging $35.77 per thousand feet.

Florida's cypress industry peaked in 1925, with 257,125,000 feet manufactured (twice around the world). Florida produced 28.5 percent of the national total, placing second to Louisiana. Florida's cypress averaged $44.18 per thousand feet, FOB mill, compared to the $40.90 national average. The best prices came in 1923, averaging $55.17 per thousand feet.

In 1912, the Florida Department of Agriculture (FDA) reported that the area of the state was 58,680 square miles, of which 4,440 were water and about 10,000 were everglades. About one out of eight acres (12.5 percent) had already been cleared for agriculture. Florida was estimated to have over 200 kinds of forest trees, more than any other state. Forty-six species of common commercial trees made up almost 97 percent of that forest. The remaining 165 species were not of much commercial importance because of scarcity, small trunk size, or poor trunk form. Florida's forest wealth was, therefore, a wealth of species rather than a wealth of wood.

In 1912, five of our most abundant species supplied 97 percent of the lumber used (but not necessarily grown) within the state. That lumber was 70.10 percent longleaf pine (*Pinus palustris*, Miller), 12.20 percent Cuban pine (*Pinus cubensis*, Griesebach), 6.34 percent loblolly pine (*Pinus taeda*, Linn.), 6.30 percent cypress (*Taxodium distichum*, Rich.), and 2.07 percent shortleaf pine (*Pinus echinata*, Miller).

Southern Florida, according to the US Weather Bureau, lies south of latitude 28°55′ (at Turtle Mound). On the peninsula, summer heat and winter cold are tempered by the Gulf of Mexico and the Atlantic Ocean. Seasons do not follow the conventional three-month division. Spring is March through May; summer, June through September; fall, October through November; and winter, December through February.

In February 1902, the Weather Bureau established a Cooperative Observer's Meteorological Station at Malabar, five miles south of Melbourne. Seasonal summaries, based on records kept prior to 1912 (before Melbourne's logging era began), told the newcomers what they could expect weather-wise.

Spring skies were mostly clear, with winds from the east-southeast and an average temperature of 72.4 degrees. The high was 98 degrees and the low 37 degrees. Eighteen days had some precipitation. Mean monthly rainfall was 2.90 inches, with a monthly high of 8.38 inches and low of 0.06 inches. The greatest rainfall within any 24-hour period was 3.26 inches on April 9, 1911.

Summer skies were mostly clear, with winds from the southeast and an average temperature of 80.9 degrees. The high was 101 degrees and the low 59 degrees. Eight days were at or above 100 degrees, with a record high of 101 degrees on August 19, 1909. More than half the total annual rainfall occurred in the summer, mostly as heavy afternoon thundershowers lasting around an hour and a half. Forty-five days had some precipitation. Mean monthly rainfall was 6.22 inches, with a monthly high of 19.96 inches and low of 1.40 inches. The greatest rainfall within any 24-hour period was 9.79 inches on September 17, 1908.

Fall skies were mostly clear, with winds from the east-northeast and an average temperature of 72.7 degrees. The high was 97 degrees and the low 32 degrees. Fifteen days had some precipitation. Mean monthly rainfall was 4.09 inches, with a monthly high of 12.62 inches and low of 0.39 inches. The greatest rainfall within any 24-hour period was 3.80 inches on October 24, 1902.

Winter skies were mostly clear, with winds from the northeast and an average temperature of 62.3 degrees. The high was 89 degrees and the low 21 degrees. Twenty-three days were at freezing or below, with a low of 21 degrees recorded on January 26, 1905. December and February were the driest months, with nighttime rains almost as likely as daytime rains. Seventeen days had some precipitation. Mean monthly rainfall was 2.11 inches, with a monthly high of 6.05 inches and low of 0.06 inches. The greatest rainfall within any 24-hour period was 3.10 inches on December 20, 1911.

Between 1887 and 1911, two tropical depressions, 18 tropical storms, and nine hurricanes passed within 100 miles of Melbourne. During the logging era (1912–1932), seven tropical storms and three hurricanes came within 100 miles.

This was the Deep South during the Jim Crow era. Race not only governed employment, but also housing, travel, and social life. In 1887, Florida enacted America's first railroad segregation law, which mandated segregated cars. In 1907, Florida law mandated segregated waiting rooms in railway stations. The terms "white," "black," "colored," and "Negro" were used openly and in books, newspapers, and even legal documents. Government and other records often contained tables with columns simply labeled "W" and either "B" or "C."

Today, it is impossible to tell who was considered "black" in the South during the early 1900s. There is no question about people of African ancestry, but the concept seems to have been applied to all dark-skinned people regardless of their ancestry. In this work, these terms appear only in their historical contexts, such as in names of organizations. Even though most of the labor force in the South was black, the term laborer, herein, covers all laborers regardless of their individual races.

East Florida had a reputation for its mosquitoes. Early Spanish explorers called the area Los Mosquitos, and the Florida Territorial Council created Mosquito County in 1824. Smudge pots, mosquito beaters made from palm fronds, heavy clothing (often lined with newspaper "armor"), and sleeping under mosquito "bars" (nets) were part of living in this paradise. Window screens helped little as the big mosquitos would push the little ones through the screen, and they in turn would unlock the door for the big ones to enter. The early natives had their own solutions, including rubbing fish guts over their skin.

One

Once Upon a Time in Melbourne

Melbourne began as a settlement on Crane Creek, one of many freshwater tributaries to the Indian River, a brackish lagoon along Florida's east-central coast. Crane Creek was about 29 miles south of Cape Canaveral. As local economy and transportation were tied to the water, Melbourne's downtown developed on the low, narrow peninsula between the Indian River and Crane Creek.

The first three settlers, freedmen from Georgia, Alabama, and Virginia, arrived in the early 1870s, coming overland along former military trails. Other settlers came overland and by boat on the Indian River. By June 1880, the village of Crane Creek had six families, 21 residents, and a US Post Office.

In July 1880, the community took the name Melbourne, honoring one of its early settlers who came from Melbourne, Australia. Many of these early settlers were English descendants. By 1885, the population was 50.

By 1910, the town had grown to 157 and had become widely known for its fine hunting and fishing, proximity to pristine ocean beaches, rich farm country, low incidence of hurricanes and, as was most of east Florida, for its mosquitoes! Melbournites had their own newspaper, post office, telegraph, and telephone. Farming, ranching, fishing, and related trades supported the local economy. There were no large industries, electricity, ice production, hospital, or municipal water and sewage systems.

Primary transportation along the coast was by the Florida East Coast Railway (FEC) or by boat on the Indian River. Local transportation was by foot, horseback, buggy, and wagon. The "horseless carriage" was just beginning to appear. "Roads" were basically sandy wagon trails.

Melbourne was, however, almost totally isolated from the interior of the state. Only a couple of sandy and almost always submerged wagon roads ran west across the six-mile-wide floodplain and virtually impenetrable marshes of the upper St. Johns River. Travelers had to either use the hand-operated ferry 11 miles west at North Indian Field or ford the river itself 17 river miles farther north at Lake Winder. No railroads or highways crossed the river south of Enterprise Junction, some 80 miles to the north. (See maps on pages 21 and 50.)

The highest point in Melbourne was the fourth-floor cupola atop the Brown House, a hotel near the FEC station. Around 1910, an unidentified photographer captured this panoramic view looking east down New Haven Avenue toward the Indian River, about a quarter of a mile away, and the barrier island (locally known as the "eastern peninsula") some 1.5 miles beyond. On a clear day, the ocean, yet another mile away, can be seen. Residential Melbourne dominates the

foreground, 25 feet above the lagoon, from the four-story Carleton Hotel, on the bluff on the left, south to the mouth of Crane Creek. The downtown business district is all but invisible on the low peninsula between Crane Creek and the Indian River, extending from the triple-roofed building at the end of New Haven Avenue south to Beaujean's boat works. (SBHS.)

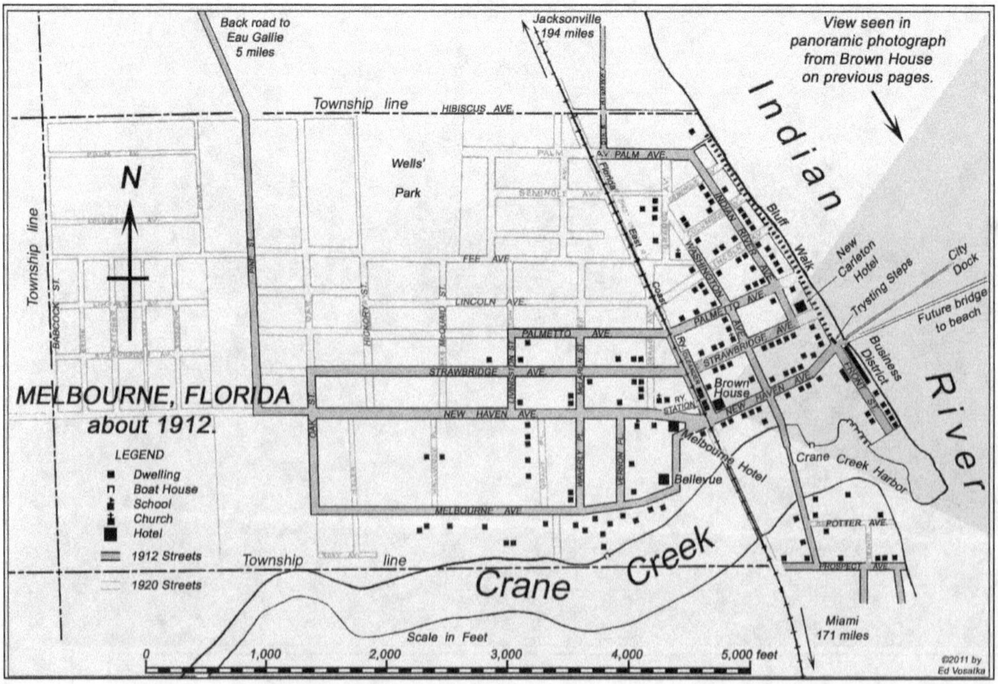

Melbourne's downtown occupied the peninsula between the Indian River lagoon and Crane Creek. This peninsula, in reality a big sandbar at the mouth of the creek, created and sheltered a naturally protected harbor. Front Street, the only main street in downtown, ran south from New Haven Avenue. Businesses on the east (left) backed up to the lagoon, several having their own docks. Fourteen men, three women, and six businesses held county/state-issued licenses during fiscal year 1912. Twelve were for merchants, three for greengrocers, two each for real estate agents, hotel managers, launch operators, barbers, and fish dealers, and one each for a druggist, liveryman, undertaker, banker, lunch stand, turpentine still, and carriage/wagon factory. The map above combines elements of several maps from January 1887 through March 1920. (Above, author's illustration; below, Ann Flotte.)

At the east end of New Haven Avenue, to the left in the previous photograph, was the municipal pier, built in 1886. The Indian River was about two miles wide at this point, with an average depth of three feet. The pier had to be 1,400 feet long in order to reach five feet of water for the steamboats. A narrow-gauge car, pushed by hand along iron rails, carried passengers and freight. The big steamers were gone by this time, pushed on south, hastening their own demise by helping build Flagler's railway. Yet, the Indian River was still a major artery of transportation. Almost everyone owned a boat, from rowboats and sailboats, to steam launches, to modern naphtha launches. There were ferryboats and trade boats. Doctors and dentists traveled the river from town to town in their floating offices. (Weona Cleveland.)

The *Mystic*, owned and operated by Capt. Albert Rogero, was a typical commercial trade boat. County-wide, 23 licenses were issued during the 1912 fiscal year for operating launches, each with a $5 state and $2.50 county fee: 15 to carry passengers or freight, 1 to carry passengers or under 15 tons of freight, 5 to carry under 15 tons freight, and 2 with no specifications listed. (SBHS.)

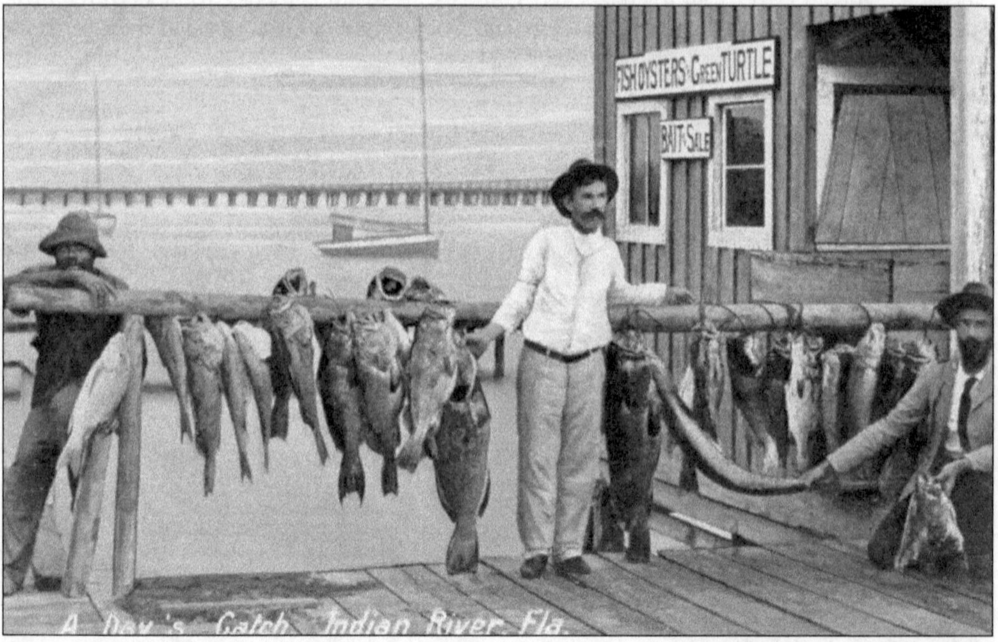

In January 1897, Commissioner J.J. Brice, of the Department of Fish and Fisheries, in a report to the US Senate, listed 106 species of fish occurring in the Indian River. Of these, 26 were deemed commercial. The 11 most abundant food-fish, in order of value, were as follows: mullet, pompano, sheepshead, sea trout, redfish, mangrove snapper, bluefish, whiting, sailor's choice, southern flounder, and silver mullet. Turtles and oysters were also abundant. (Author's collection.)

Ferryboats provided a much-needed service, carrying passengers, groceries, freight, and mail across the Indian River to the residents living on the eastern peninsula. There was no bridge, and there would not be one until May 1921. The *Ida Mae* (launched c. 1898), *Jessie B* (launched c. 1903), and the *Atlantic* (launched in 1918), owned and operated by Rufus and Claude Beaujean, made trips daily except Sundays. (Weona Cleveland.)

The ocean beach was one of Melbourne's tourist attractions, a popular place for picnicking, surf-bathing (swimming), fishing, and even camping. After taking a ferryboat across the Indian River to Melbourne Beach, people could board a motor-powered railcar and ride to the beach. Anywhere else, they faced a mile-long trek through the palmettos and other dense vegetation. Going to the beach could become an all-day affair. (SBHS.)

The "trysting steps" at the north end of Front Street, built in 1886, provided tourists and residents an easy climb up the 25-foot river bluff to the hotel and residential areas. At the top was the Bluff Walk, a scenic path running seven blocks north along the edge of the bluff. The Carleton Hotel and many residences fronted this path overlooking the Indian River. (Author's collection.)

The 24-room Carleton Hotel opened in February 1887. On December 27, 1903, a fire, originating in a cottage to the north, was driven southward by strong winds. Within two hours, two cottages and the Idylewild and Carleton hotels were consumed. Rebuilt, it reopened November 1, 1904, as the New Carleton Hotel. Daily rates were $2 and up, and amenities included sanitary plumbing, furnace heat, gaslights, and private baths. (Author's collection.)

Heading west from the foot of the pier, animal-drawn and motor vehicles had to take New Haven Avenue to reach the upper village residences, hotels, and railway station. On the left are the home and garden of Annie (Richard W.) Goode and family. On the right is Madame Campbell's Bazaar. (Author's collection.)

New Haven Avenue became a tree-lined residential street at the top of the bluff. The first cross-street was Washington Avenue, Melbourne's main north–south street and the road from Jacksonville to Miami. Downtown buildings and Washington Avenue are visible in the distance, looking back east along New Haven Avenue. (Weona Cleveland.)

Travel by road was limited to going north or south along the Indian River and somewhat inland at various communities. Local sandy roads were being covered with shells dug from the adjacent Indian mounds. The road from Jacksonville to Titusville had been completely hard-shelled since October 1912. The first automobile was seen in Eau Gallie on November 17, 1906. In 1914, seven private-use licenses and 16 commercial "for hire" licenses were issued in the county. Licenses cost $5, all of which benefited the state. A $2.50 county charge was added later that year. Melbourne had an eight mile-per-hour speed limit, which included mules and horses, and fines ranged from $5 to $25. (Above, author's collection; below, Ann Flotte.)

A wagon road (old Indian trail) going west from Melbourne led to a public ferry, similar to this, crossing the upper St. Johns at North Indian Field. In December 1900, the Brevard County commissioners let a $125 contract to have this road upgraded to fourth class, and in August 1903, it was upgraded to third class. In February 1911, it was declared a public road. (See maps on pages 21 and 50.) (Author's collection.)

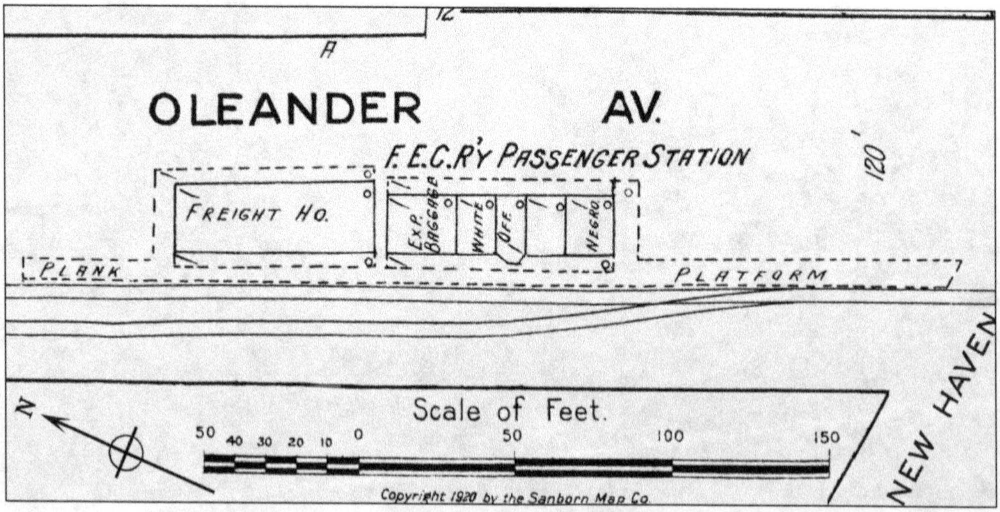

A bit farther west along New Haven Avenue were the Brown House, the railway depot, and the single-track main line. Henry Flagler's Jacksonville, St. Augustine & Indian River Railway (renamed the Florida East Coast Railway in September 1895) reached Melbourne in mid-1893, and the first passenger train arrived on October 2. By timetable, Jacksonville was a 194-mile, nine-hour trip, plus a 23-hour overnight layover in Palatka. By April 1912, there were three daily all-Pullman passenger trains, and travel to Jacksonville was only 8.25 hours regular and 6.75 hours express. Fares were calculated at 3¢ per mile, so the Melbourne to Jacksonville ticket would cost $5.85. Melbourne's station was the original yellow and white wood structure but was remodeled in 1907 to provide the segregated waiting rooms mandated by Florida law. (Above, author's collection; below, Weona Cleveland.)

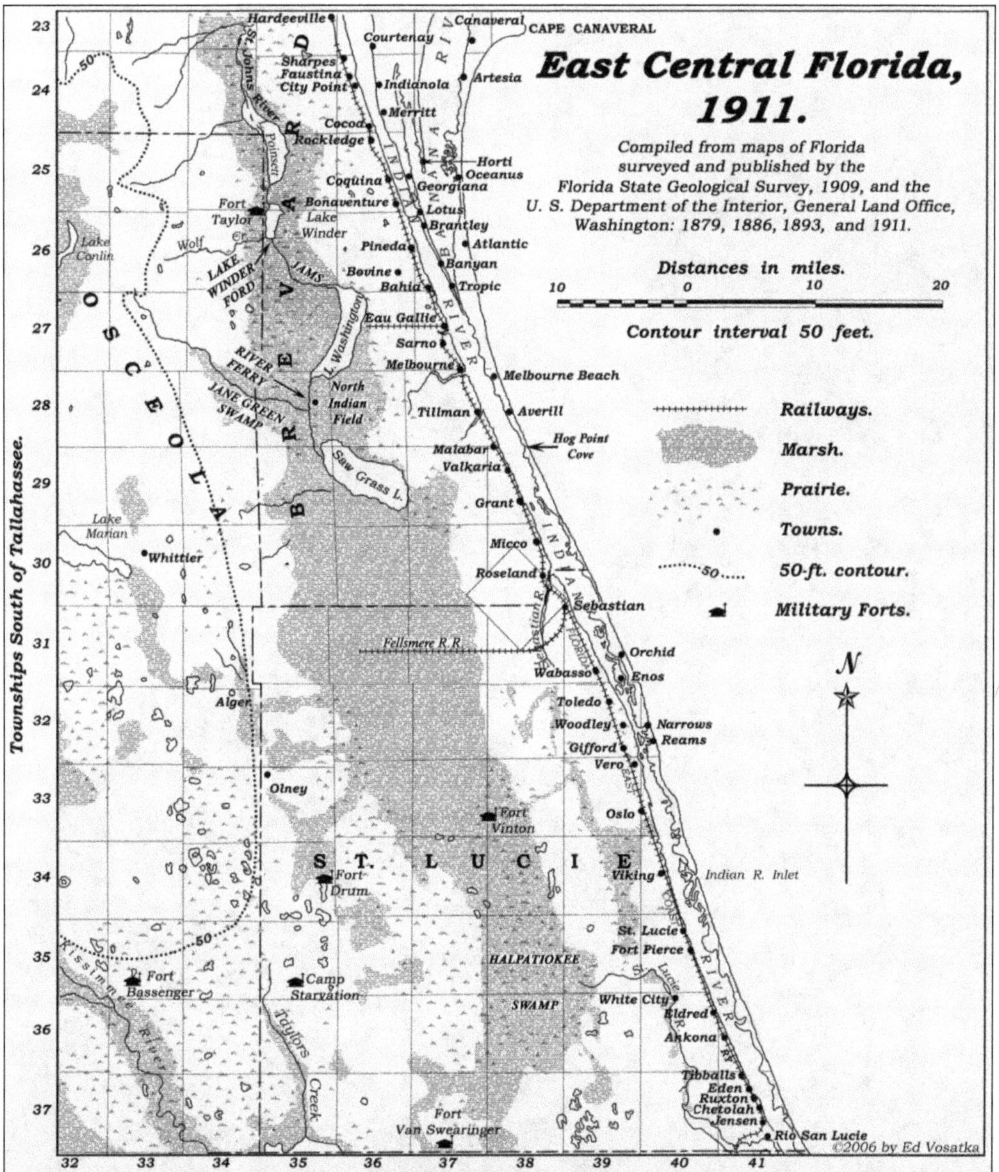

The virtually impenetrable marshes of the upper St. Johns were still the formidable barriers first described by GLO surveyors. "The remainder of this township is utterly impracticable" was noted near Lake Washington on two 1848 township maps. "Impreticticable marsh & swamp, unsurveyed" was printed over five townships west of Roseland and Wabasso on the 1879 *Map of Florida*. The nearest practical crossings were some 80 miles to the north. Getting to Kissimmee, only 53 miles due west of Melbourne, entailed a 128-mile journey via the FEC and Atlantic Coast Line (ACL) railways: a 6.5-hour ride, including changing trains at both Titusville and Enterprise. By road, via New Smyrna, Deland, and Orlando, was 153 miles. At the state speed limit of 10 miles per hour (5 on curves), if the sandy roads would permit that speed, an automobile trip would be at least 15 hours driving time, plus overnight lodging. By horse, walking three to four miles per hour and covering 20 miles a day before resting, took about a week. A better route was sorely needed. (Author's illustration.)

George Washington Hopkins (born November 8, 1844) was an experienced lumberman from Manistee County, Michigan, with extensive background in surveying, brick manufacturing, and railway construction and operation. He founded the Village of Bear Lake in 1874, where he established the Bear Lake Lumber Mill, the first livery stable in northern Manistee County, and the first roller gristmill in the upper part of lower Michigan. He also organized the Michigan Hardwood Lumber Association. He and his wife, Ella, started wintering in Florida for his health around 1890, spending time in Kissimmee and St. Cloud. In 1902, he started buying tracts of land at St. Cloud. From 1902 through 1907, he purchased some 100,000 acres (156 square miles) of choice country land around a small ranching community called Crab or Crabgrass, paying about $1.04 per acre. This land lay west of Eau Gallie and Melbourne, across the St. Johns astride the Osceola-Brevard county line. He named his ranch Ocean Prairie, after the vast, windswept grasslands stretching for miles along the St. Johns from marsh to tree line. (Helen Fairman.)

Two

A Town Called Hopkins

George Hopkins now owned 58,000 acres of virgin timber, 240 million feet of long-leaf yellow pine on the uplands, and 225 million feet of bald cypress in the magnificent Jane Green, Penna-Wah, and Wolf Creek swamps. He considered logging in mid-1903 and selected a site for his mill town along the newly chartered Central Florida & Indian River Railroad, a planned 150-mile trunk line from a 35-foot-deep ocean port at Hog Point Cove near Malabar (see map on page 21) to the phosphate and citrus regions around Bartow. Hopkins built their roadbed where it crossed his ranch at his own expense. The CF&IR was surveyed and partially graded but was never finished.

In 1911, with his timber now valued in excess of $2 million, he turned to the Clark Brothers, a prominent lumber firm from Alexandria, Louisiana. Charles and Milton Clark came to Melbourne to scout locations for a mill. September saw them in Eau Gallie, site of the now-defunct Edge and Company sawmill and its standard-gauge logging railway to Lake Washington. By late October, they decided to build south of Crane Creek, just outside Melbourne's corporate limit, in an area known as "South Melbourne." A partnership was brokered, and the two Clark families moved to Melbourne.

Charles Clark personally purchased 157 acres of pine flatwoods in November and December for $1,802. This property, 15 to 20 feet above the lagoon and next to the FEC, offered adequate timber for construction, a large expanse of flat land for the mill and lumber storage, a natural water supply fed by a series of ponds along the coastal ridge to the south, and good natural drainage north into Crane Creek. Being near, but not within, Melbourne also allowed them to use local laborers without having to pay city taxes.

The USDA, in its *Soil Survey of the Indian River Area, Florida* of 1915, described flatwoods as primarily longleaf and slash pine with an understory of saw palmetto, gallberry, and wire grass (three-awn). Its sandy soil had a high water table with a hardpan layer about three feet down. Land of this character generally ran $10 to $60 per acre, depending on location and drainage.

The Union Cypress mill and power plant were structural steel buildings with corrugated steel siding and roofing. Material and equipment may have come from the old Titusville Lumber Company mill that was demolished in December 1910. The three-story mill was about 50 by 155 feet, the power plant about 42 by 75 feet. The open space under the mill contained power transmission equipment: engines, shafting, loaders, conveyors, and similar equipment. The second floor housed the lumber manufacturing equipment: saws, edgers, trimmers, transfers, and accessories. The third floor was the saw-filing room. Footers for columns were set about three feet into the fine sand soil. Machinery foundations, susceptible to heavy loads and vibrations, were poured around wood pilings driven down eight or so feet to coquina bedrock (see page 122). Construction began in March 1911, and sawing commenced October 28. (Above, author's collection; below, Bob Rowe.)

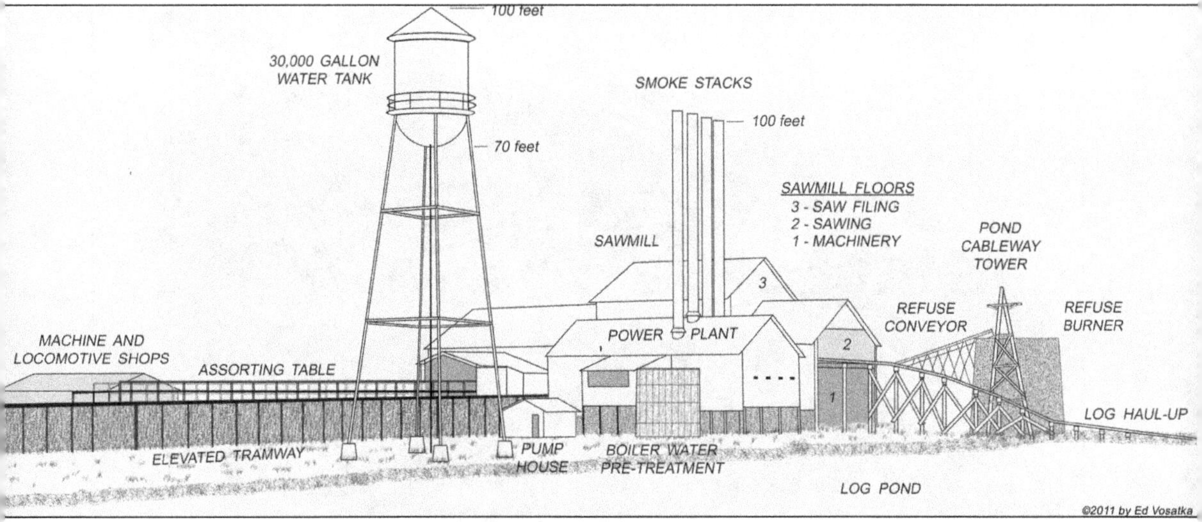

Sawmill plants and operations were fairly standardized. Cypress mills were recognizable by their large drying yards. The five-acre log pond served as fire protection as well as for log storage. Two dedicated fire pumps were on six-inch wells of 125- and 150-foot depths. Because steam engines and pumps take time to reach optimal speed, fire pumps were kept running in alternation so one was always ready for an emergency. An elevated water tank provided gravity-fed fire protection and a water supply for the community. In the power plant were four boilers and their auxiliary equipment. Boiler feed water was treated to remove impurities, which could cause damage. Here also were the main engines, one driving the mill and the other driving the electric generator. All electrical switching equipment was in this building. The mill was designed to cut 38,460 feet of lumber a day (enough to build a bookshelf from Melbourne to Malabar-Valkaria). Bandsaw blades, moving about 125 miles per hour, dulled quickly depending on the wood and were exchanged and sharpened three times a day. (Author's illustration.)

Steam was supplied by five Casey & Hedges (Chattanooga) boilers. These 1921 catalog cuts show the patented design that simplified field construction and repairs by inexperienced mechanics. Each 150-horsepower, horizontal-return-tube (HRT) boiler was six feet in diameter and 18 feet long, with a 125- to 150-pounds-per-square-inch working pressure. A steel casing covered the entire boiler setting. Inside this casing was a layer of ordinary brick and a layer of refractory brick. Highly efficient Dutch-oven furnaces generated intense heat and were ideal for the lowest grades of fuel and refuse impossible to burn in ordinary furnaces (sawdust, hoggings, spent tan, bagasse, etc.). Having complete combustion and sparks and unconsumed fuel removed before they entered the boiler flues made these furnaces practically smokeless. A boiler purchased new by the Union Cypress in November 1913 cost $1,889, complete. (Both, Hagley Museum & Library.)

A two-cylinder steam engine, located in the power plant, "pulled" the mill machinery. It was connected by a broad leather belt to a wheel on the main drive shaft under the mill. The engine "ran over," meaning the top of the flywheel moved away from the cylinders. This caused the upper part of the belt to run slack and have more surface area in contact with the wheels at each end. (CabinCreekCDs.)

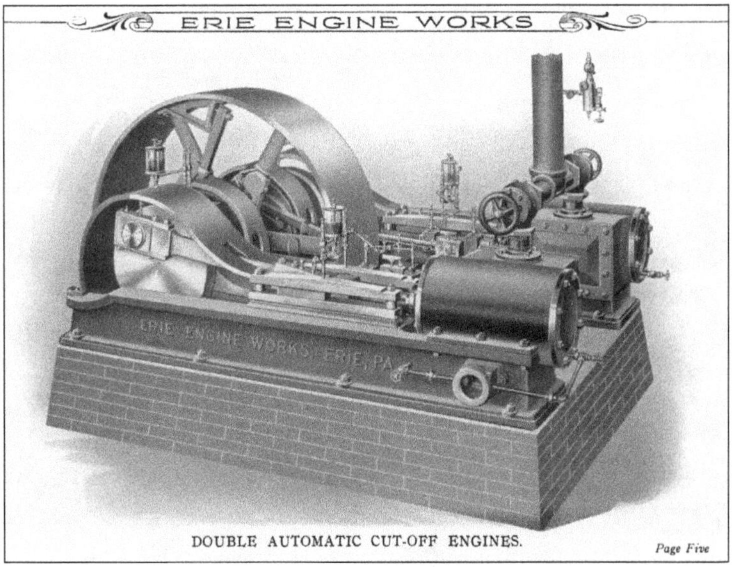

A General Electric alternator and high-speed Erie City Iron Works engine, similar to these, were installed in February 1913. They produced a 2,300-volt, three-phase, 75-kilowatt current for the mill. The town of Hopkins received electricity by March 23. Melbourne's city-owned distribution system was activated January 2, 1915, and the city paid the Union Cypress an average $175 per month for electricity. The city in turn charged its customers 15¢ per kilowatt, with a $1.50 minimum. (Author's collection.)

South of the mill was the heavy timber loading dock. "Timber" meant wood sawed four inches by four inches and over in size, which was shipped green. Timbers were cut to length and trimmed to exact size ordered, then carried outside by conveyor and stored. Once a carload had accumulated, it was loaded into a flatcar or gondola, such as this one from the Southern Railway. (Author's collection.)

The drying yards were southeast of the mill. Common grades of softwood lumber were often air-seasoned, while higher grades were kiln-dried. Lumber was first inspected and sorted by size and grade on the "assorting table" (foreground), then placed on carts and pushed by hand along tracks on an elevated tramway (center) to the drying yards (background). There, the lumber was stacked by size to dry and be stored until shipped. (FHS.)

Union Cypress No. 1, a used locomotive rebuilt by Southern Iron and Equipment Company (SI&E) of Atlanta, was purchased on November 25, 1911. She was built about 1905 by Baldwin Locomotive Works of Philadelphia for Tennessee's Chattanooga and Montlake Railway. Her original builder's number is unknown, but she was probably a class 10 1/2 C. Locomotive No. 2, below, also a SI&E rebuild, was purchased May 31, 1912. She was built in April 1871 as Lehigh Valley 107, an 0-6-0, in their Hazleton, Pennsylvania shops. Hazleton rebuilt her in December 1886 as LV 3233. Acquired by SI&E in October 1905, she was refurbished and then sold to Bartow Lumber Company of Cartersville, Georgia, on May 2, 1907. Back again at SI&E, she was converted to a 2-6-0. (Both, SI&E Collection, Archives Center, National Museum of American History, Smithsonian Institution.)

To better understand the layout of the sawmill complex and surrounding residential areas, one must jump ahead 16 years to the year 1928, when this landmark aerial photograph was taken. In 1919, the US Coast and Geodetic Survey, in cooperation with the US Army Air Service, began to study the feasibility of using aerial photography for coastal mapping. This study successfully proved that aerial photography could replace plane-table surveying, producing more detail at a lower cost. Missions were flown in Florida (1926–1928) for coastal chart revision using a Fairchild T-2 four-lens camera aboard a Loening AP 0A1B, No. 27-319. Neighboring images overlapped along the flight line, permitting stereoscopic viewing. Flight 699 recorded 70 images between Malabar and Bellwood in April 1928. This image of South Melbourne, with north to the left, covers an area about 2,436 by 3,073 feet. It was enlarged in 1978 from NOAA copy negatives F-682 and F-683. (Author's collection.)

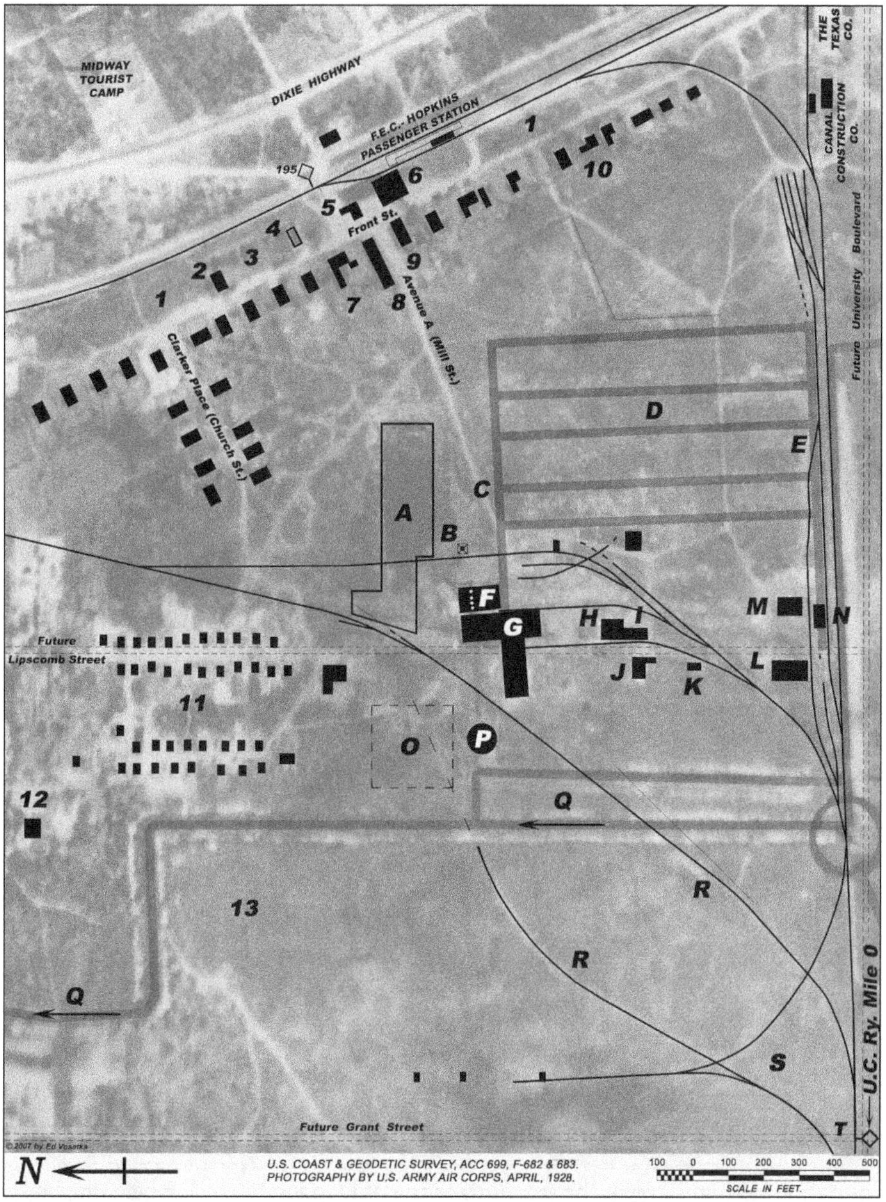

Many buildings and other structures of the mill complex and South Melbourne can be seen in this aerial photograph. In the residential area are (1) housing along Front Street for single and married white employees, (2) Union Church (1915), (3) park and playground, (4) swimming pool, (5) company office, (6) commissary, US Post Office, and doctor's office, (7) boardinghouse for single white men, (8) Crescent Theatre (1914), (9) Dr. Martin's "sanatorium," (10) Hopkins public school (1916), (11) housing for laborers, (12) Melbourne Vocational School (1919), and (13) Stone Cemetery for Union Cypress employees (1922). In the industrial area are the (A) log pond, (B) water tank, (C) elevated tramway, (D) drying yards, (E) loading platform, (F) power plant, (G) sawmill, (H) machine shop, (I) locomotive shop, (J) mill supply, (K) hay storage, (L) dry kiln, (M) planing mill, (N) timber sizer on loading dock, (O) ice plant (1917–1920), (P) trash burner, (Q) drainage canals to Crane Creek, (R) railway spurs, (S) railway wye for turning trains, and (T) the railway main line to Deer Park. (Author's collection.)

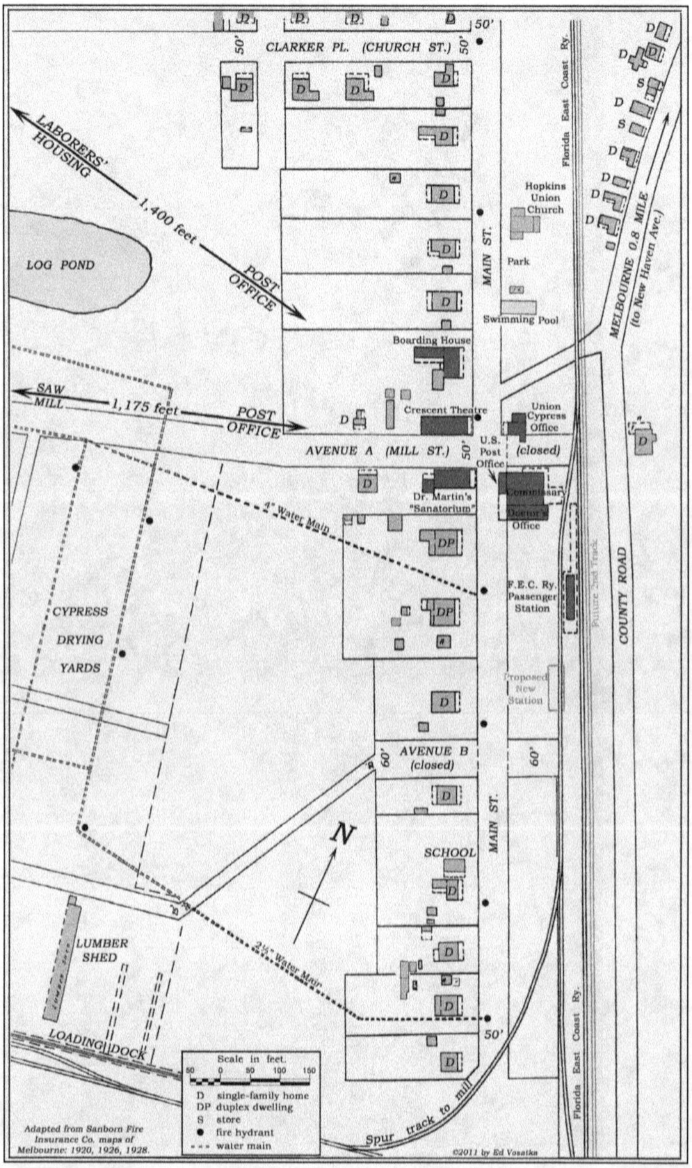

In a typical Southern mill town, the sawmill was the only industry, the lumber company owned the dwellings, and administration and police functions were controlled by company officials. The Union Cypress town had no official name. "Downtown" consisted of the company office, commissary, post office, doctor's office, sanatorium, theater, swimming pool, park, church, and FEC Railway station. Along Main Street was a two-story boardinghouse for single white male workers (rent unknown) and 11 single-family and two duplex dwellings for white mill officials and their families (rent $10 per month). Housing was racially segregated. A separate residential area for laborers was northwest of the mill. Here, 40 single-story "shanties for laborers" were clustered along Centre (Lipscomb) and Race Streets (rent $1 per month). These nonindustrial buildings were typically rough pine western frame construction with box sills supported about two feet above the ground on concrete footers. Roofs were shingled. Light and water were provided from the sawmill plant, but there was no sewage system other than outhouses and septic tanks. Many other employees lived in Melbourne, Tillman, or in the country. (Author's collection.)

The company office (left) and commissary (right) were located at the "front" of the town, along Main Street and the FEC Railway. The Union Cypress Company was incorporated on December 4, 1911, with $400,000 authorized capital stock, divided into 4,000 shares at $100 each. The board of directors included D. Frank Clark, president; Charles H. Clark, vice president; and George Hopkins's son-in-law Clifton C. Burgess, secretary treasurer. On July 8, 1912, C.H. Clark resold all the property he had purchased in the company's name to the company at cost. George Hopkins, who held the largest amount of stock, 1,970 shares, was the fourth member of the board. In 1913, he became company president and Milton Clark became vice president. The announcement below appeared in the *Melbourne Times* on July 16, 1913. (Above, Patricia Hook; below, *Melbourne Times*.)

Company currency was in the form of metal coins and paper script. The coins were a patented design (No. 632,938) with a brass rim and an aluminum insert, made by S.D. Childs and Company, Chicago. They were the same sizes and denominations as regular currency and were used to pay for food, dry goods, sundries, and other items at the commissary. Some Melbourne merchants also accepted them at face value. Many laborers were hired by company recruiters traveling through Georgia, Florida, and the Carolinas, with most hires coming from Georgia. A total of 300 men were on the payroll in December 1912. By March 1914, there would be 600. Everyone worked a 10-hour day, six-day week, with an hour off at noon. The mill was closed on Sundays. Employees were paid either daily, for their previous day's work, or monthly, based on a daily rate. Typical wages in the Florida lumber industry can be found in *Wages and Hours of Labor in the Lumber, Millwork and Furniture Industries, 1915* from the US Bureau of Labor Statistics, Bulletin 225, February 1918. (Author's collection.)

The commissary carried foods, dry goods, sundries, and other household items. Prices were the same for both employees and nonemployees. The first business license obtained by Union Cypress (April 1912) was for the commissary, with a $5,000 stock of merchandise. Inventory grew to $10,000 for fiscal year 1914 and $15,000 for fiscal year 1918. The commissary was in the 50-by-60-foot main building, the post office was in the front addition, and the company doctor's office was on the right. Just out of view to the right was the Hopkins-FEC Railway passenger station. The gentleman below, on the left, is Frank S. Meyer, commissary manager and postmaster. The others are unidentified. (Above, Patricia Hook; below, Sterling Hawks.)

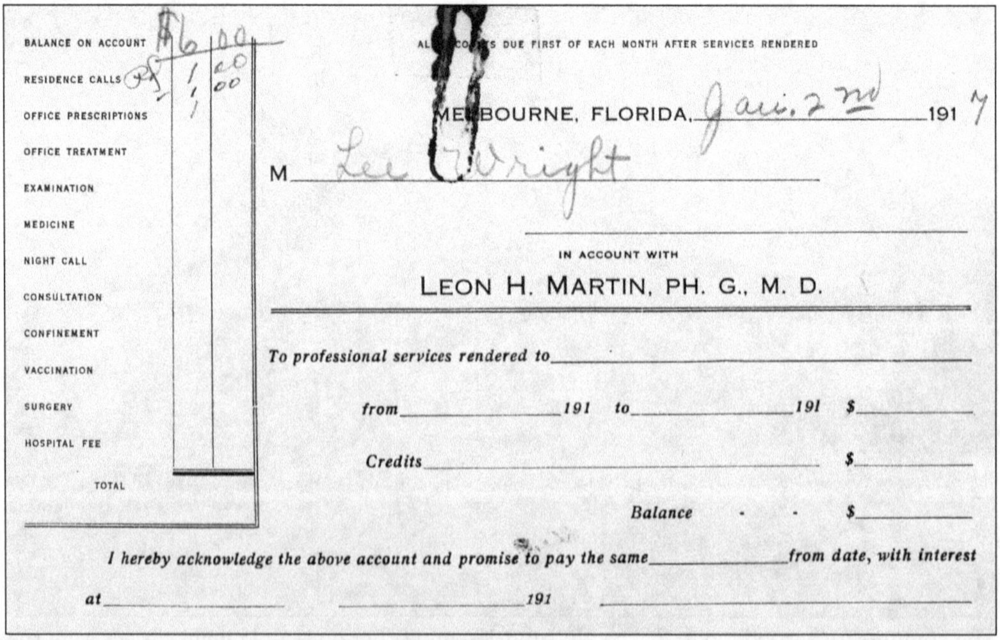

BALANCE ON ACCOUNT	$6.00
RESIDENCE CALLS	1.00
	1.00
OFFICE PRESCRIPTIONS	
OFFICE TREATMENT	
EXAMINATION	
MEDICINE	
NIGHT CALL	
CONSULTATION	
CONFINEMENT	
VACCINATION	
SURGERY	
HOSPITAL FEE	
TOTAL	

ALL ACCOUNTS DUE FIRST OF EACH MONTH AFTER SERVICES RENDERED

MELBOURNE, FLORIDA, Jan. 2nd 1917

M Lee Wright

IN ACCOUNT WITH

LEON H. MARTIN, PH. G., M. D.

To professional services rendered to

from ____ *191* *to* ____ *191* *$* ____

Credits ____ *$* ____

Balance - - *$* ____

I hereby acknowledge the above account and promise to pay the same ____ *from date, with interest*

at ____ ____ *191* ____

Melbourne did not have a hospital in 1912 and wound not until 1926. The nearest one was in St. Augustine, some six hours by train, with only one morning and two afternoon northbound trains a day. Dr. Leon H. Martin, 22, arrived from Enterprise, Alabama, in early August 1912. Holding degrees in both pharmacy and medicine (PhGMD), he served as the company physician and surgeon, working out of his office next to the commissary. Medical services were equally assessed at $1 per month for single men and $1.50 per month for married men, regardless of race or the number of children. This fee was taken out of the employee's pay. Dr. Martin began specializing in surgery while here. (Above, author's collection; left, Bob Rowe.)

SCHEDULE REFERRED TO HEREIN

X-Ray outfit, including table, tubes, lead glass shield, fluroscope, dark room outfit, plates, paper, etc. etc.

Microscope and accessories, including centrifuge, stains, slides, cover glasses, glass bowl cover, carrying box, etc. etc. Tables, an adjustable enamel top instrument table, operating table, etc. etc.

Roll and flat top desks.

Typewriter desk, two typewriters, swinging desk stand, etc. etc. Office chairs and twelve (12) desk chairs, cushions for chairs, etc. etc. Card index desk and letter files, instrument cabinet.

All surgical and obstretic instruments, surgical, obstretric and medicine bags. All drugs. Operating gowns, caps, gloves, towels and racks.

Two (2) alcohol stoves, lamps.

One (1) battery, electric treatment lamp, one (1) urea outfit, infusion set, stomach pump, two (2) colon tubes, one (1) atomizer; masks, choloroform and ether.

Automatic soap dispenser.

Basis, trays, sterilizers, largs and small, one (1) large and one (1) small sterilizer, glass funnel, pitcher, bottles and jars, scales and weights. water sprinkler, three (3) sets crutches, one (1) observation stand.

All instruments and accessories, accumulations, etc. etc. contained and now in my hospital at Hopkins, Florida.

WITNESS: J. W. Martin M.D. (Seal)

George F. Eschbach (N.P.SEAL)

Frank L. Bills

Filed & Recorded 4th June 1917

J. F. Mitchell Clerk Ct Ct By [illegible] DC

Dr. L.H. Martin practiced in Hopkins until the end of May 1917. When he left to enlist in the Army Medical Reserve Corps, he sold his equipment for $2,000 to Dr. J.W. Martin, no relation, of Palatka, who replaced him. Dr. J.W. Martin moved his office into the "Martin Sanatorium," and the old office became a storeroom for the commissary. In late September 1917, George Kempfer became the first patient to enter the Hopkins Hospital, when he had surgery to remove his adenoids. Dr. J.W. Martin remained in Hopkins until mid-1920, when he left to go to the booming sawmill town of Osceola, just north of Lake Harney. (Brevard County Chattel Mortgage Book Y.)

Melbourne's first moving picture house, the Crescent Theatre, opened in the south part of town in January 1914. It had a stage and screen with seating on the main floor and racially divided balcony. Bob Rowe was manager and projectionist, and Richard Henley sold tickets. Race determined seating and admission (35¢ or 15¢). Films came first to Cocoa, then on to South Melbourne by train for a Saturday night showing. (Bob Rowe.)

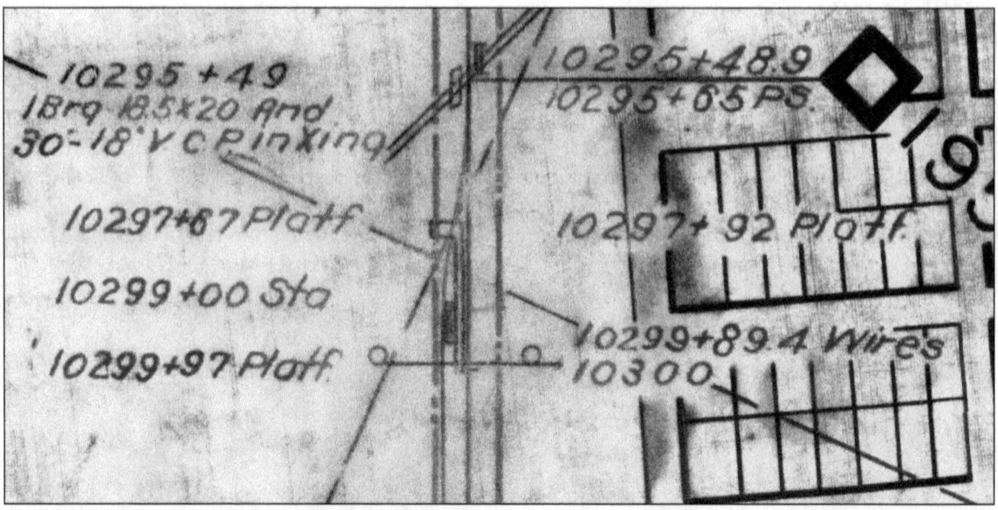

The Florida East Coast Railway passenger and freight station for South Melbourne was in operation by June 16, 1914, although it would not receive an official name for another three weeks. It was a typical FEC-designed wooden building, about 16 by 64 feet, with a 232-foot platform set on the town side (west) of the main line, between it and a side track. Three passenger trains stopped daily going each way. (Author's collection.)

UNION CYPRESS COMPANY

MANUFACTURERS OF

LUMBER
SIDING
CEILING
FLOORING
MOULDING

Red Cypress

LATH
PILING
SHINGLES
CROSS TIES
RAILROAD TIMBER

GEORGE W. HOPKINS, PRESIDENT
J. M. CLARK, VICE PRES. C. C. BURGESS, SECY-TREAS.

OFFICE AND MILL AT
HOPKINS, FLORIDA

July 3rd 1917

$1131.00

The town officially became Hopkins in July 1914. Although it bore his name, G.W. Hopkins did not name the town. It was merely the fourth name Frank S. Meyer submitted to the Post Office Department, which had rejected his first three choices. A post office was established on September 23, 1914, and opened November 1, with Frank Meyer as postmaster. Services available were money orders, Western Union telegraph, and Southern Express shipping. There was no home delivery service and would not be until 1926. First-class postage was 1¢ at first, then was raised to 2¢ on November 3, 1917, and 3¢ on July 1, 1919. (Both, author's collection.)

The Union Church of Hopkins sat on land purchased by church trustees from the Union Cypress in January 1915. This 25-by-50-foot building had two wings and a bell tower. Rev. Charles E. Hauck, pastor at the Congregational Church in Melbourne, also served as pastor at Hopkins. For Easter 1918, "The church was decorated in green and white with an arch of lilies across the entire rostrum. Two tables on either side were banked with white water lilies and the piano was covered with oleanders," according to the *Melbourne Times*, April 3, 1918. When the Hopkins church closed in early 1926, its bell was preserved by the Congregational Church, and it is now located at that church's former location on the corner of US 1 and Strawbridge Avenue. (Both, Sterling Hawks.)

All company houses were basically the same style of construction. Single white workers lived in this two-story boardinghouse on Main Street, which still stands virtually unchanged. Originally, there was also a detached "summer kitchen." Exterior walls were one-by-eight-inch vertical board and batt. Interior walls throughout were one-by-eight-inch horizontal shiplap. Some boards still bear marks from the original 48-inch circular saw. The roof was originally covered with a composite asphalt material. On the first floor were the kitchen, a combination parlor/dining room, and six boarding rooms. Upstairs were six boarding rooms and two rooms for the manager and his family. Doors were solid, and entrances also had screen doors. Windows were glazed, two over two, single-hung, and had screens. Each room could hold two bunk beds, or four men. In the mid-1920s, the building was expanded and became a tourist hotel, the Hopkins House. (Above, author's photograph; below, author's illustration.)

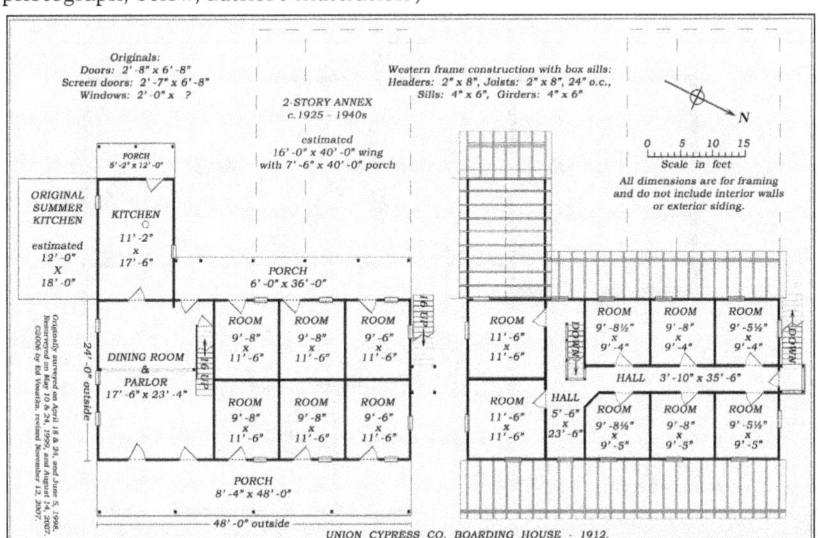

In 1912, the company built 13 rental houses along Main Street for married white employees and their families. These were typically single-story bungalows with four to five rooms plus a kitchen. Each was about 28 by 28 feet with a 14-by-12-foot kitchen wing and a full-width front porch. A high-pitched roof enclosed a large attic, which could be finished as an additional half-story. By early 1920, a total of 21 such residences had been be built. Engineer John Madden and his family lived just north of the boardinghouse. Pictured in 1930 from left to right are John, Rossie, Sallie, Myrtice, and Sallie's brother Pershing. The floor plan depicts the bungalow seen on page 120, which still exists. (Above, Myrtice Madden Floyd; below, author's illustration.)

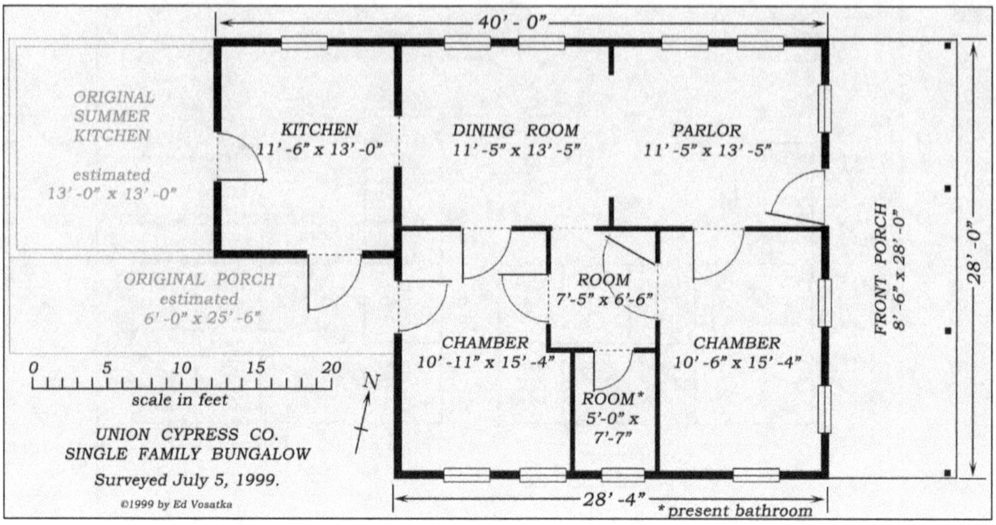

South of the commissary were two duplexes for married white employees and their families. Both were 1.5 stories, about 43 feet wide by 46 feet long, with a kitchen wing and full-width front porch. Commissary manager Frank Meyer and his family lived in the house shown above, and William H. Kempfer and his family lived in the house shown below. Kempfer graduated from University of Michigan's School of Forestry in 1906 and married George Hopkins's daughter Agnes, also a 1906 UM graduate, on September 25, 1907. Kempfer worked seven years for the US Forestry Service, including the US Forest Products Laboratory at University of Wisconsin, and published many technical articles. The Kempfers moved to Hopkins from Los Angeles in September 1913 and later moved out to Deer Park. (Above, Sterling Hawks; below, Katherine Doran.)

A swimming pool was located on Main Street just north of the company office. Standing in the pool is Richard W. Henley, who came to Melbourne from Indiana with his parents in the late 1890s. He began working as the company bookkeeper in October 1912. In late 1920, he took C.C. Burgess's place as secretary. The crossbucks in the background were at the FEC/Jernigan Avenue grade crossing. North of the pool, between it and the church, was a park, seen below. Playing in the park are, from left to right, John, Rossie and Myrtice Madden and Helen and Ernest Spikes. John's favorite toy was his toy pistol. (Both, Sallie Madden.)

The Hopkinses had a home in Kissimmee and, in 1903, they started coming to Eau Gallie for the Christmas holidays, staying with friends. In May 1907, George Hopkins hired Eau Gallie contractors C.G. Redstone & Son to build a large houseboat, about 16 by 55 feet and drawing about two feet of water. She was a comfortable and roomy double-decker, with accommodations for eight guests plus crew quarters. Onboard were all possible conveniences, including a toilet, ice chest, pantry, and water tanks. She had a powerful Strellinger gasoline engine, purchased direct from the factory in Detroit. Christened *Wolverine*, for Hopkins's home state, she launched in mid-December 1907. The *Wolverine* became the family's winter home, and they cruised the Indian River many times, often entertaining guests on board. On one evening cruise, about 20 guests were on board. After the *Wolverine* became unseaworthy in early 1910, Hopkins purchased a lot on the Eau Gallie River, had her hauled ashore, and converted into a house. The Hopkinses spent their next 15 winters in their new residence, also named "Wolverine." (Helen Fairman.)

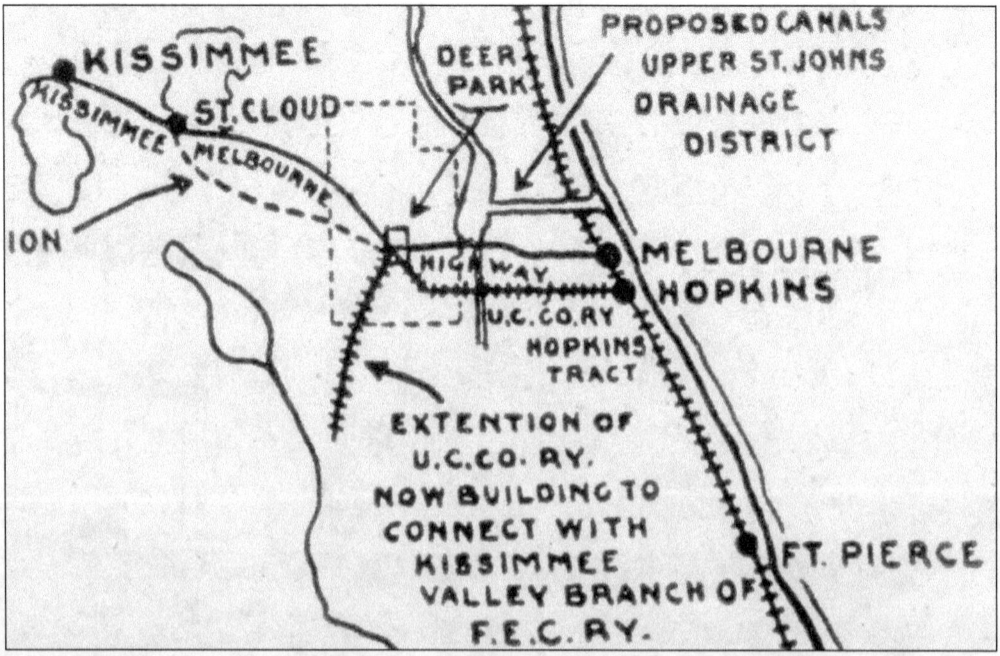

Logging railways were generally not surveyed and constructed to the same precision as main line railways. The shortest and cheapest route was the best. Vice Pres. C.H. Clark stated his philosophy was that "it (cost) but little more to build a real road than it (did) to build a makeshift." He planned to extend the Union Cypress to Narcoossee and tie it in to central Florida's rail systems. (Author's collection.)

Union Cypress rail came from FEC's Atlantic and Western Division (New Smyrna to Orange City Junction), which was built between 1883 and 1886 as the Blue Spring, Orange City & Atlantic Rail Road (BSOC&A). Rail brands tell where and when a rail was rolled: these by Scranton Steel in November 1885 and Pennsylvania Steel in 1883. Rail also came from Bethlehem Steel (1883) and Lackawanna Iron & Coal (1884). These were all Bessemer steel mills located in Pennsylvania. (Author's photograph.)

Three

Red Rails in the Sunset

Along with the mill property, Charles Clark purchased a 50-foot right-of-way from the mill to the edge of Hopkins ranch—10 miles for a total of $601. It appears to have followed the abandoned CF&IR Railroad grade.

After the roadbed was surveyed, workers with 30 mules "swamped" (cleared) all brush and trees within the right-of-way. This width was sufficient to keep "down timber" from falling on the tracks during windstorms and to help control wildfires started by locomotive sparks. Stumps and roots that could interfere with the roadbed were "grubbed" (completely removed).

Mule teams pulling drag scrapers formed the roadbed, pulling sand from the sides toward the middle, and forming a raised center with drainage ditches on either side, about 20 feet apart. Mules were the animals of choice in the South. Compared to horses, mules were better adapted for hard labor in hot weather, able to stand rougher treatment, less excitable, and more agile on rough ground. They ate less, seldom overate, and performed more labor on poorer feed. In the South, clearing a right-of-way by contract labor averaged $25 per acre, about $151.50 per mile for the Union Cypress. Grading and ditching a roadbed ranged from $3 to $4 per 100 feet.

Ties were cut from yellow pine owned by Richard W. Goode (350 acres for $2,500) about two miles west of downtown as well as from trees within the right-of-way. Typically eight feet long and square-cut to an eight-inch width and six-inch height, ties weighed about 145 pounds green and 100 pounds air dried. Untreated pine ties were expected to last four to six years.

In December, the UC leased from FEC "18 miles of 50-pound relaying rail" (50 pounds per yard, used) and other track fixtures for $36,450 plus interest. The same amount of new rail, at 1910 mill prices, would have cost $39,600. The rail was about four inches high and manufactured in standard 30-foot lengths. One mile of track required at least 2,640 ties, 352 rails, 704 joint bars, 1,408 sets of fasteners, and 10,560 spikes.

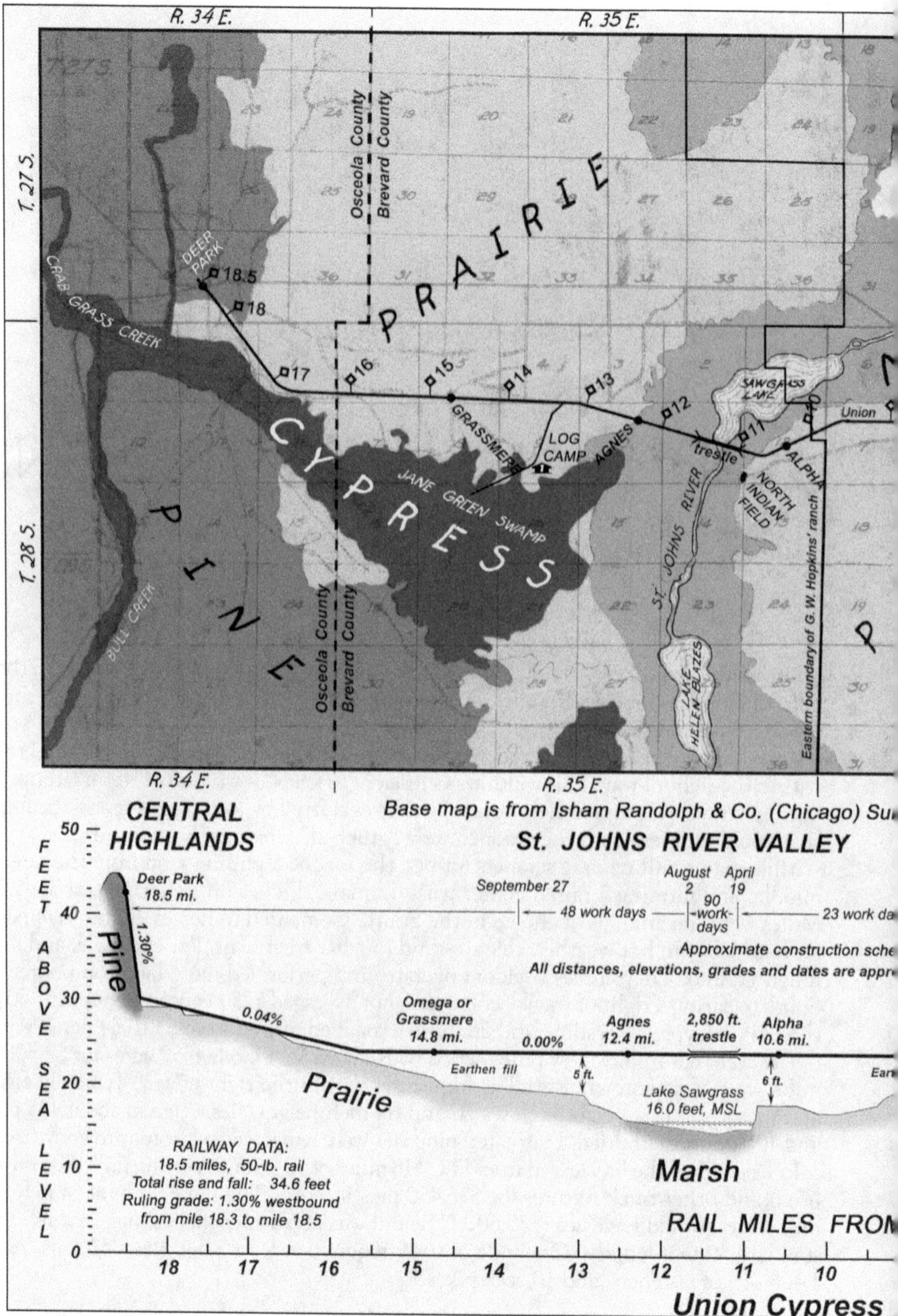

FEC added a main line siding in South Melbourne in January 1912. A spur line to the mill allowed rail, locomotives, rolling stock, and other heavy equipment to be brought in. On February 6, George Hopkins wrote, "six miles of rail now laid and nine miles of grade." In late March, the Clarks provided a free excursion for about 160 Melbournites over the first eight miles of their railway. Workers reached the St. Johns in mid-April and crossed it by early August. By late

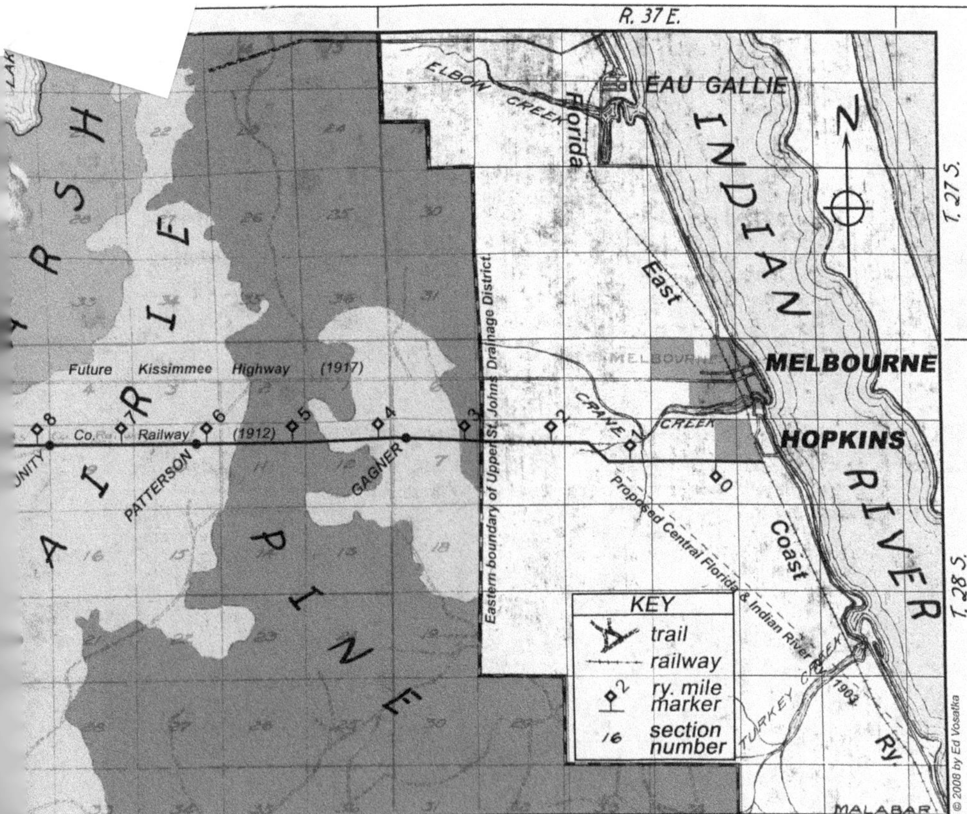

f the Upper St. Johns Drainage District, June 1920.

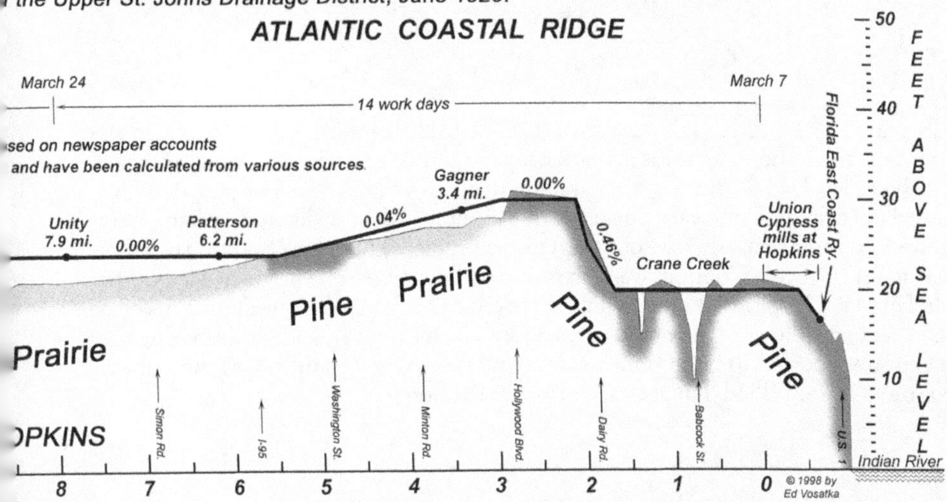

way, 1912.

September, the Clarks were offering free public excursions to the Jane Green log camp, leaving the mill at 9:00 a.m. and returning at noon. Enough timber was cut and brought to the mill to allow sawing to begin on October 28. Tracks reached Deer Park ("Platt Place") in May 1913. (Author's illustration.)

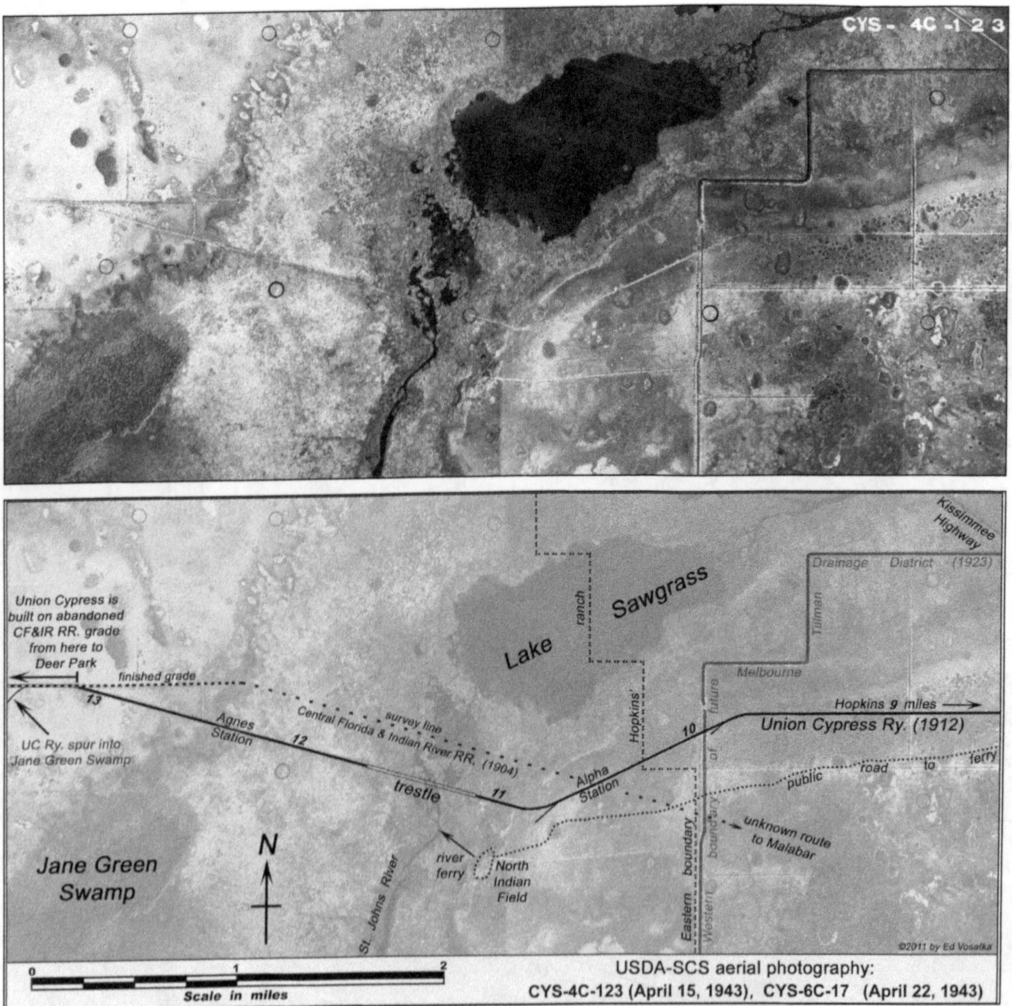

Crossing the St. Johns river and marsh was an expensive undertaking. The soil (muck and peat layered with sand) had a low load-bearing capacity and was frequently flooded for long periods. Using a Marion floating dipper dredge (one-yard bucket), crews built a 15-foot wide sand embankment across the narrowest (1.5-mile) marsh just south of Lake Sawgrass. They then bridged the river. Meanwhile, Brevard and Osceola Counties were planning a wagon/auto road between Melbourne and St. Cloud. In February, Hopkins offered to help them cross the St. Johns by widening his own embankment to 30 feet and allowing their road to run alongside his railway, with both sharing the trestle. But the counties decided to build their road north of Sawgrass, counting on the railway embankment to shield it from flood damages. (Author's collection.)

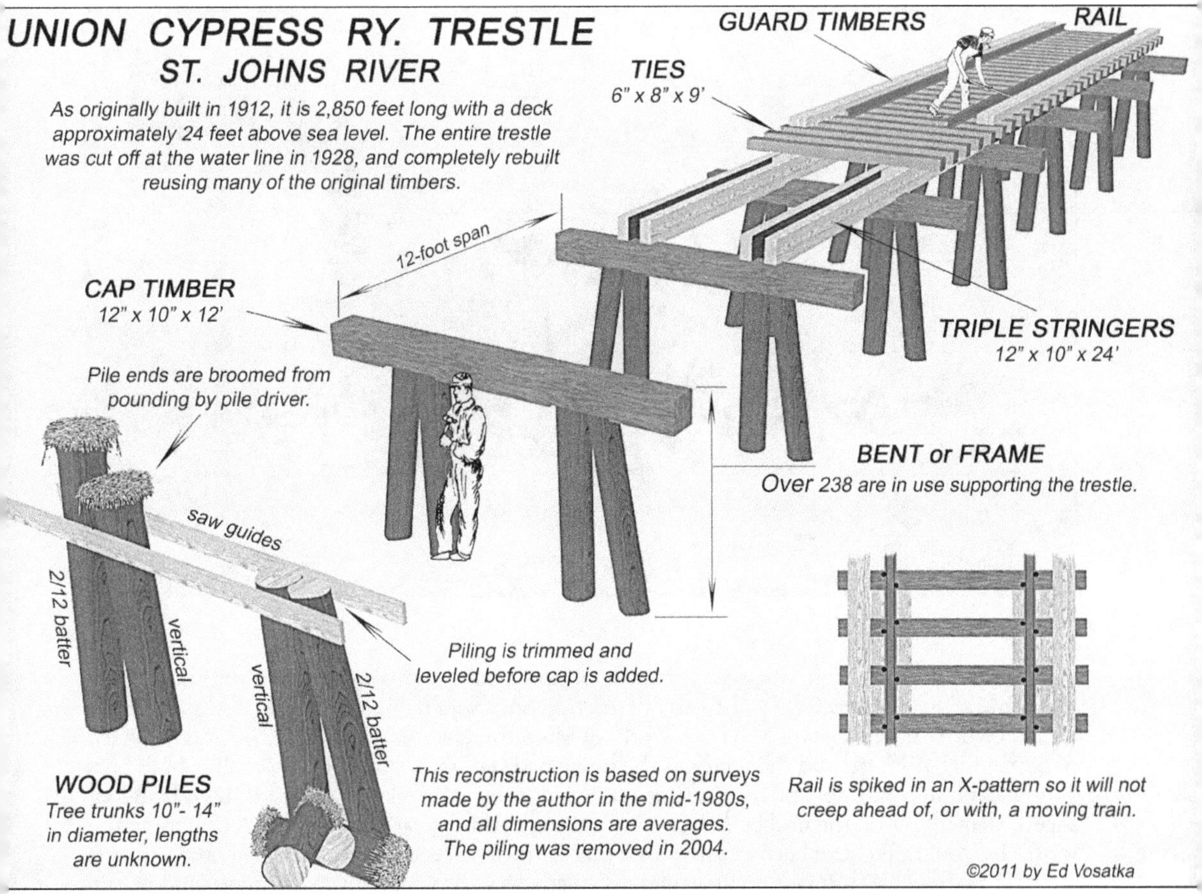

This trestle was primarily yellow pine. With a railway pile driver, workers could drive four piles in 30 minutes. Trimming them, adding the cap, stringers, tie, and rail usually took another 1.5 hours. Ideally, working six 10-hour days per week, construction could be completed in 41 days; however, floods and other challenges pushed it over 3.5 months. Untreated, these timbers could last around 10 years. (Author's collection.)

A logging train was often an odd assortment of home-built and commercial rolling stock. On the point is No. 3, a new Glover 2-6-0 (construction number c/n 14206) built on contract for the Union Cypress by the Glover Machine Works of Marietta, Georgia, costing $5,500 and delivered in September 1913. Behind the tender is the homemade combination car, or combine, built on a 30-foot car frame. Its sides and roof were of 2.5-inch corrugated steel in standard 26-inch-by-8-foot sheets, the same as on the mill buildings. Given the era and the car's design and placement in the train, this could well have been a "Jim Crow" car, having two segregated passenger compartments separated by a central freight compartment. Passengers sat on rough wood benches and put up with smoke and flying debris kicked up by the locomotive. (Bob Rowe.)

About a mile west of the mill are the headwaters of Crane Creek, called "cathead" because of the Florida panthers (*Puma concolor coryi*, Bangs) living there. Unable to roar, their scream is the best-known sound from their repertoire of low growls, chirps, hisses, whistles, and purrs. A large pale brown cat with a long, crooked tail, the males grow to seven or eight feet (including tail) and 100 to 160 pounds, the females to six feet and 60 to 100 pounds. Elsewhere, they are called catamounts, mountain cats, mountain lions, or pumas. Their preferred diet is white-tailed deer and hog. Adult Florida white-tail deer are about 36 inches tall and typically weigh 95 (doe) to 125 pounds (buck). (Above, Florida State Archives; below, author's collection.)

After leaving the mill, the first 20 minutes of travel was through an open forest of slash and longleaf pine interspersed with grassy sloughs, ponds, and swamps. This land was flat and poorly drained, and the water table was high. Water often stood in low places for long periods during rainy seasons, occasionally interfering with the trains. Unaware of crossing the Atlantic Coastal Ridge, only about 12 feet higher than the mill, one would pass turpentine laborers gathering pine gum collected in "boxes" cut into the bases of the trees—an old method now being replaced by cups hanging on the sides of the trees. The firm of Powell & Shave owned or leased turpentine rights to 30,000 acres in the Melbourne area. Their still was located just southeast of Hopkins. (Both, author's collection.)

Thunderstorms are frequent and severe in the southeastern coastal regions. Lightning may ignite a dry tree or the decaying leaves and branches at its base. The Forest Service, in its *Lightning in Relation to Forest Fires*, Forest Service Bulletin 111, conservatively estimated in September 1912 that lightning struck 30,000 trees each year, with longleaf pine being hit the most. Lightning, however, was second to sparks from locomotives in setting forest fires. (Author's photograph.)

The 1913 hunting season ran from November 20, 1913, to February 20, 1914, for deer, squirrel, wild turkey gobblers, quail, turtledoves, swans, geese, brant, ducks, rails, coots, mud hens, sandpipers, curlews, snipe, and plover. By state law, daily limits were "1 buck, 2 gobblers, 20 quail or 25 birds of any other species." For this hunting season, the limits were "3 bucks, 5 gobblers or 500 of any other game bird species" (gallinules, shore birds, surfbirds, woodcock, tattlers, grouse, and pheasants). No other animals were protected by law. (Ann Flotte.)

A broad, nearly level prairie, predominantly sand cordgrass with scattered cabbage palms, bordered the St. Johns marsh. Ancient shell mounds (c. 4,000 years old) left by native tribes formed isolated knolls with clumps of palms, scattered pine, saw palmetto, and wiregrass. Scattered depressions held wetland grasses, flags, and lilies. The marsh was a virtually impenetrable mix of sawgrass, maidencane, flags, sedges, and scattered buttonbush on a nearly level peat and muck soil. The river originated in the Halpatiokee swamps west of Fort Pierce and flowed north to Jacksonville (see map on page 21). In this flat upper basin, the river could flow either way, depending on wind and rain. Islands of floating vegetation, the "jams," blocked the river north of Lake Washington, curtailing the northward flow of water. During rainy seasons, the river backed up, forming a huge, shallow lake, and the upper marshes also drained south into the Okeechobee. (Both, author's collection.)

Water hyacinths, a tropical South American native, were widely cultivated as ornamentals in Europe and America. In 1897, the USDA (*The Water Hyacinth*, USDA Bulletin 18) told how they came to be in the St. Johns. Around 1890, a person in Edgewater, near Palatka, in cleaning a pond, tossed hyacinths into the river. Passing settlers saw them, thought they were pretty, and took them home. In October 1894, hyacinths appeared in the Ocklawaha River, 14 miles downstream from Silver Springs. People were not the only culprits though. Riverboats and currents also carried them around. Stockmen were said to have brought them to the upper St. Johns for cattle food. For a few years, Florida's hyacinths were limited to the St. John, its tributaries, and a few inland lakes. No one knew these beautiful, light-blue flowers would quickly become "Florida's million-dollar weed." (Both, author's collection.)

Lake Washington, considered the head of river navigation, 276 miles above Jacksonville, was one of the most productive fisheries in the Southeast. The jams blocked upstream-bound boats, so few knew of this fishing paradise until November 18, 1869, when the 75-foot steamer *Pansoffkee* entered the lake on an expedition to judge the feasibility of a navigation canal between the lake and the Indian River. The Palatka *Times-Herald* of March 6, 1896, told of the steamer *Alligator* bringing back "1,000 pounds of pickled fish, representing . . . over 600 of the finny tribe. The largest fish captured was a ten and a half pound bass . . . an eighteen-inch moccasin snake was taken from the stomach of an eight and one-half pound trout." Above, from left to right, Dan Martin, Pete Olden, Bob Rowe, and Johnny Hawkins are about to try their luck. (Above, Bob Rowe; below, author's collection.)

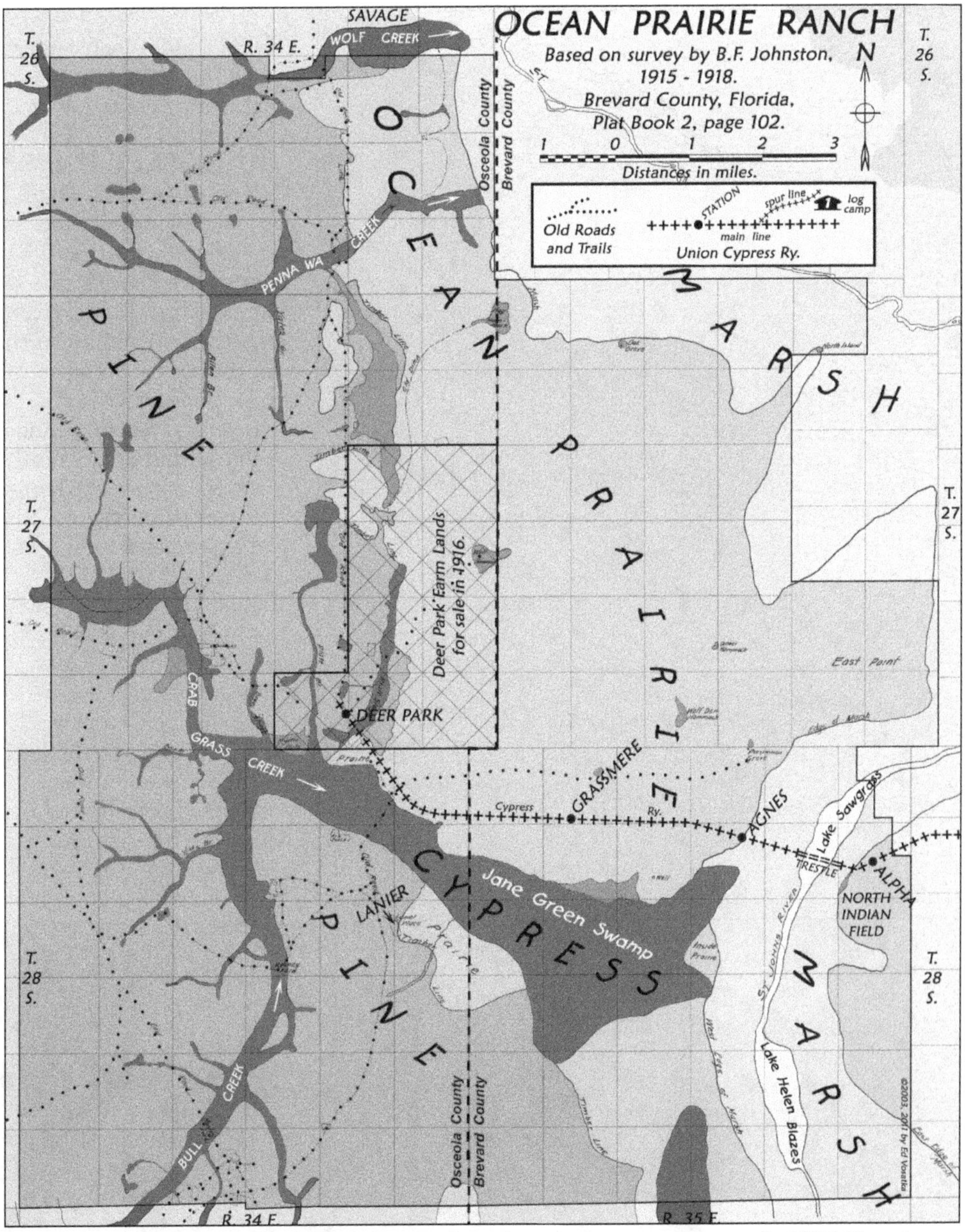

Beyond the marshes was the "Ocean Prairie," a name dating to at least early 1897. Farming was not practical here because of poor access, but good grazing was. Intending to upgrade the native scrub cattle, Hopkins began experimenting with different strains of cattle, grasses, and forage crops to determine what would be best for this climate. The ranch was completely fenced in, contrary to prevailing free-range practices. Originally called the Ocean Prairie Ranch, the name Deer Park Stock Farm first appeared around February 1904 and was incorporated January 22, 1916. The Hopkins herd grew from 650 head in October 1902 to 2,000 by July 1913 and 12,000 by May 1916. Some 11,000 head of improved scrub cattle were shipped to the Agar Packing Company of Chicago in June and July 1917. Ten 30-car trains carried them over the Union Cypress to Hopkins and thence to Chicago over FEC and other lines. Better-grade, tick-resistant Aberdeen-Angus and Brahma breeds took their place. (Author's illustration.)

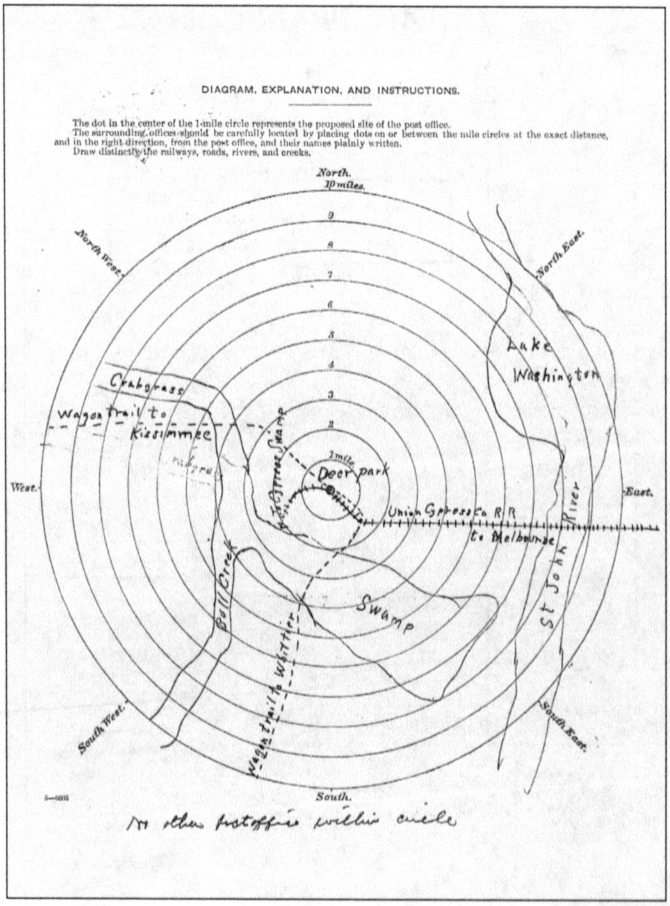

The Hopkins ranch centered around an older cattle settlement called Crab or Crabb, which had a post office from July 3, 1896, to January 18, 1899. The surrounding countryside was still known by its settlements: Platt Place, Sumner Place, Robb Place, and others. Ten miles north was the Savage Settlement on Wolf Creek. Other settlements were on Penna-Wah Creek, six miles north (see map on page 59). Lanier, 2.5 miles southwest of Crab, had a post office from May 16, 1899 to July 31, 1914. William H. Kempfer drew this map to complete his application for a new post office, to be called Deer Park. His application was submitted February 18, 1914, and the post office was established September 23, 1914, with him as postmaster. (Left, National Archives; below, author's collection.)

At Deer Park, the firm of Hopkins & Kempfer ran a naval stores industry, general store, and other commercial businesses. The town also had a logging camp. The gum naval stores industry involved the production of turpentine and rosin from pine gum, or resin. Slash and longleaf pine were tapped in the United States. Trees were "cupped" between November 1 and March 15 (see page 54, bottom), and yearly operations were coordinated in advance with logging operations. Deer Park had a 25-barrel copper still (50 gallons per barrel) that could distill three "charges" per day, producing two to five gallons of turpentine per charge. Many of the turpentine workers, here attending a baptism in Crabgrass Creek around 1919, were recruited from other states. (Above, Carolyn Kempfer; below Ray Bush.)

DEER PARK FARM LANDS

Deer Park, Florida

Being a Part of the Hopkins Tract Comprising 104,000 Acres in Brevard and Osceola Counties

George Hopkins and William Kempfer decided in mid-1916 to sell some 6,000 acres (see map on page 59) for settlement, ranching, and farming and to publish this 16-page, 8-by-9-inch illustrated booklet. They had 494 half-acre in-town lots (60 by 135 feet) and 80 within a proposed town expansion, for $50 and up. Beyond were about 5,500 acres of practically clear prairie for $30 to $50 per acre. Its dark sandy loam, well filled with humus, was ready for the plow, and its heavy growth of wire grass offered excellent grazing. Deer Park was just over an hour by train from Melbourne and soon would be connected to both Melbourne and Kissimmee by road. The land sloped gently towards the St. Johns. Shallow, parallel ditches, being cut with this Austin ditcher (US Patent 1,467,828), provide sufficient drainage. (Both, author's collection.)

By May 1916, a total of 45 houses had been erected and more were under construction. The town had two general stores, a post office, Union church and Sunday school, a two-story hotel, orange and grapefruit groves, truck gardens, two good artesian wells, a doctor, two carpenters, a turpentine still, logging camp, and plenty of work for everyone. William H. Kempfer (right) was postmaster and ran the general stores, each carrying about $3,000 in merchandise and enjoying a rapid turnover from a good trade with folks in town as well as out on the farms. Union Cypress paymaster Bob Rowe (left) left the company in 1914 and moved to live and farm at Deer Park. A.B. McMullen (center) represented the Gulf Life Insurance Company. (Right, Weona Cleveland; below, author's collection.)

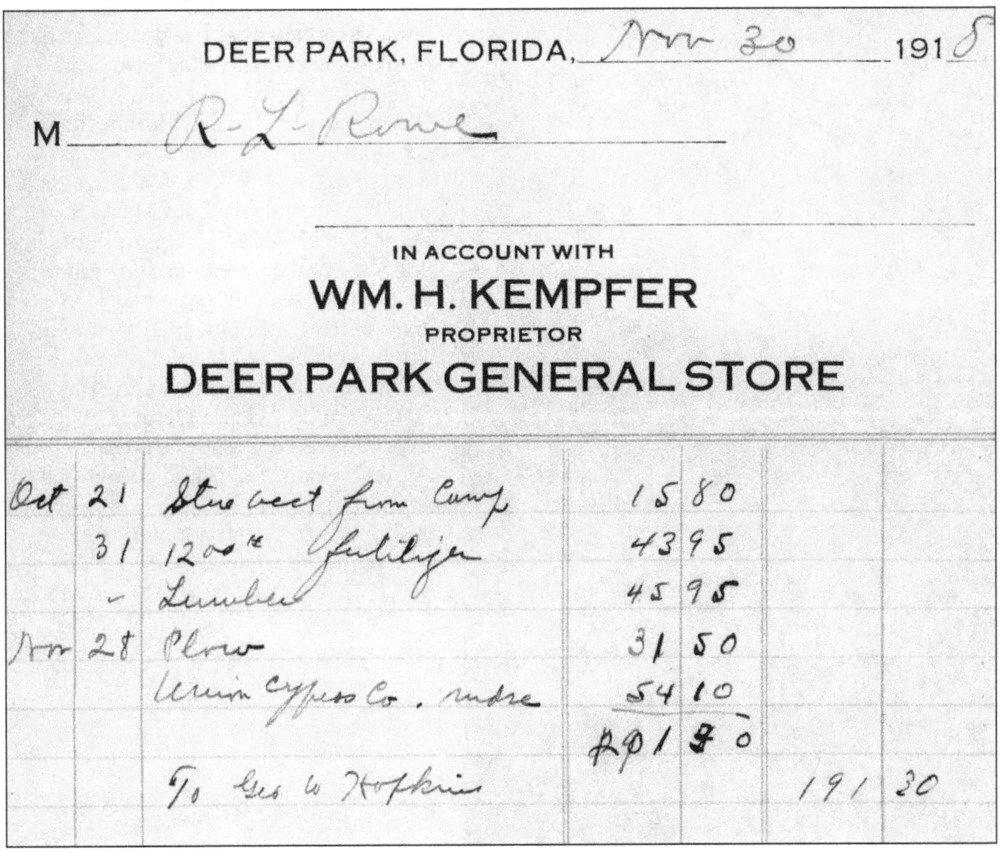

DEER PARK, FLORIDA, Nov 30 1918

M R-L-Rowe

IN ACCOUNT WITH

WM. H. KEMPFER

PROPRIETOR

DEER PARK GENERAL STORE

Oct	21	Stew beef from Camp	15	80		
	31	1200# fertilizer	43	95		
	"	Lumber	45	95		
Nov	28	Plow	31	50		
		Union Cypress Co. mdse	54	10		
			$191	30		
		To Geo W Hopkins			191	30

For over 30 years, range cattlemen—pioneer farmers themselves—have produced good crops of corn, vegetables, and citrus in this section. "In most cases, commercial fertilizer has not been used at all, yet excellent yields have been secured," reported the *Kissimmee Valley Gazette* on October 22, 1915. In the first six months of 1917, over 5,000 acres were sold, tracts of 40 to 640 acres each, mostly to actual farmers. Experienced potato growers from Hastings, Bunnell, and Kissimmee were breaking ground for the next season's estimated 1,000 acres of Irish potatoes. In 1918, a total 11 of the larger growers collectively had 535 acres in potatoes, including the 50 acres of J.M. Clark, pictured above in June 1918, looking southwest from Agnes Station. Sea Island cotton (left) was also tried, but it failed in the second year due to soil depletion. (Above, *Melbourne Times*; left, Cliff McNulty.)

By 1916, over 20 families called Deer Park home. Thirty pupils attended Public School No. 31, which sat on two acres in town owned by the Osceola County Board of Public Instruction. Robert Donegan was supervisor, and Diadamia C. Wheeler was this year's teacher, with a salary of $40 per month. School years varied in length; this one ran from August 7, 1915, through early May 1916. Deer Park did not get a black school (to be School No. 32) until 1918. (Author's collection.)

This unidentified group of people is enjoying an outing in 1919, picnicking on an abandoned trestle somewhere near Deer Park. The town had a Woodman of the World youth camp, civic league, and other social facilities befitting a small village. The Melbourne Woman's Club has held several meetings and picnics here. (Ray Bush.)

Vol. V JANUARY, 1917 No. 12

THE DISSTON CRUCIBLE

Cypress Logs

This is the kind of cypress cut by the Union Cypress Company of Hopkins, Fla. It is red cypress and some of the logs measure 7 feet 6 inches in diameter.

The company uses Disston saws.

191

The Union Cypress is featured in this advertisement in the January 1917 issue of the *Disston Crucible*. Henry Disston and Sons, of Philadelphia, was founded in 1843 and had offices worldwide. Their annual production, as listed in 1894, was 50,000 circular saws; 200,000 crosscut, mill, mulay, and drag saws; 72,000 hack and butcher saws; 516,000 compass, keyhole, and web saws; and 50,400 brick trowels. They also produced 30,000 hand saws per week and 12,000 files per day. The original photographs can be seen on page 80. (*Disston Crucible*.)

Four

JANE GREEN'S CYPRESS

Jane Green's legacy began in 1850s central Florida. Jane and her sister's husband ran off together, abandoning his wife and baby. After passing through Fort Christmas, they settled on a high mound in the cypress swamp and built a palmetto shanty. Two weeks later, Jane's three brothers showed up. Evidently, Jane would not leave her man, and they died together and were buried together in the hollow of a great cypress log, according to the *St. Andrew's Buoy* of December 9, 1897.

George Hopkins's 225 million feet of cypress grew primarily in the Jane Green, Penna-Wah, and Wolf Creek swamps (see map on page 73), enough for a bookshelf from Melbourne to Las Vegas and back. Under a contract dated December 11, 1911, the company would cut and manufacture all merchantable timber, standing or fallen, on Hopkins's ranch, but he would be paid only for cypress and pine (at $4 and $1 per 1,000 feet, respectively). Cypress had to be at least 12 inches at the small end of the first 16-foot log, pine had to be at least 12 inches on the stump, and all had to be cut as near to ground as practicable, leaving as little waste as possible.

Cypress grows in a narrow strip of coastal and riverine swamps from northern North Carolina around Florida to western Mississippi and up the Mississippi River to southern Arkansas. Although an evergreen, southern cypress (*Taxodium distichum*, Rich.) sheds its leaves annually, hence the name "bald cypress." Pond cypress (*Taxodium ascendens*, Brong.) is a slow-growing, stunted variety found in pine-barren ponds and highly acidic soils. The term "pecky," or "peggy," refers to any cypress that has been hollowed out by a fungus.

The Southern Cypress (USDA Bulletin 272, September 27, 1915) described cypress by trade names: "yellow"—amber or light orange-brown from deep St. Johns River swamps; "red"—dark reddish from lower Mississippi delta and other southern regions; "white"—light colored from the northern Atlantic coast and Mississippi River; and "black"—dark cypress from Gulf regions. "Red" and "black" could be interchangeable. To a logger, however, "black" meant heavy wood that would not float, while "white" meant light wood that would.

No. 4 was a new Baldwin 4-6-0, built on contract for the Union Cypress and delivered in June 1916. She cost $10,975, for which the Union Cypress paid $4,975 down in cash with two promissory notes. She was a class 10-26 D 331 (c/n 43,393) wood-burner with 56/50-inch drivers and 16-by-24-inch cylinders. The 4,000-gallon tender was equipped with a 20-foot suction hose to allow the crew to draw water from lakes and streams. Posing on the "pilot" are, from left to right, an unidentified girl; Alma Olden, wife of engineer Dave Olden; their son "D.A." Olden; Lila Rowe, wife of timekeeper Bob Rowe; and Leafy Langford, wife of engineer Tom Langford. (Left, Bob Rowe; below, Pennsylvania Historical and Museum Commission, Railroad Museum of Pennsylvania, Baldwin Locomotive Works Collection, negative 5919. Reprinted by permission.)

Union Cypress No. 4 was featured in this advertisement in the September 14, 1917, issue of *Railway Age Gazette*. The Standard Steel Works Company of Philadelphia began as the Standard Steel Works, incorporated in 1875. In 1892, they began manufacturing built-up steel tired wheels for engine trucks, coaches, tenders, and industrial purposes. In 1895, they developed a steel-tired wheel for which the tire was bolted to the wheel center. This design (US patent 589,356) was so successful that it was adopted by the Master Car Builders' Association in June 1912 as the standard MCB type of tire fastening. The Standard Steel Works Company was affiliated with the Baldwin Locomotive Works. (Author's collection.)

Cypress could be logged year-round, except during periods of high water. It was felled by crews in two-man teams using crosscut saws 6 to 6.5 feet long, with perforated lance teeth arranged in sets of four. Steel wedges, driven into the deepening cut behind the saw, kept it from becoming pinched. "Spring boards," inserted into small notches in the tree, allowed the fallers to saw above the large, swollen tree base. Saws were oiled with kerosene, which also removed accumulated tree resin. Resin left on the blade would cause it to bind. Loggers and their families stayed in camps in the woods. The first Union Cypress camp was in the Jane Green. Another, west of Deer Park, had 25 portable houses, a commissary, kitchen, and other buildings, all of which were portable on railway cars. Later, a camp would be established southwest of Deer Park. (Both, author's collection.)

The huge base of a cypress reflects its massive root system. Six to 10 strong roots descend and spread, providing a mushroom-shaped anchorage for the tree. A superficial system of large lateral roots spreads and branches outward near the surface, forming an interwoven mat stabilizing and supporting the trees in the soft soil. Lumbermen used these root mats to support their railroads within swamps. (Author's collection.)

Loggers "deaden" cypress 6 to 12 months before felling by chopping a deep, encircling girdle through the sapwood into the heartwood. As the tree dries out, it loses about half its weight. A 16-foot log, three feet across, would drop from 14 to 7 tons. Dried logs float, their toughened sapwood makes them easier to hook, and their heartwood does not swell and bind the saws. (Florida Photographic Collection.)

Spur tracks left the main line at intervals, leading to logging "sets" within the Jane Green. They were temporary and cheap, yet they had to support a fully loaded train over soft, often inundated ground. These tracks were on crib foundations, where ties are laid across two long, parallel stringers. Each stringer lines up under a rail, and is drift-bolted to numerous shorter logs placed crosswise below them (see page 81, top). Track could also be supported on a bed of closely spaced poles running lengthwise under the road with logging refuse as ballast (dunnage road), 16-to-20-foot poles laid across the roadbed between the ties (corduroy road), or spiking rail directly onto 10-to-12-foot poles laid across the roadbed in lieu of ties (pole road). (Both, author's collection.)

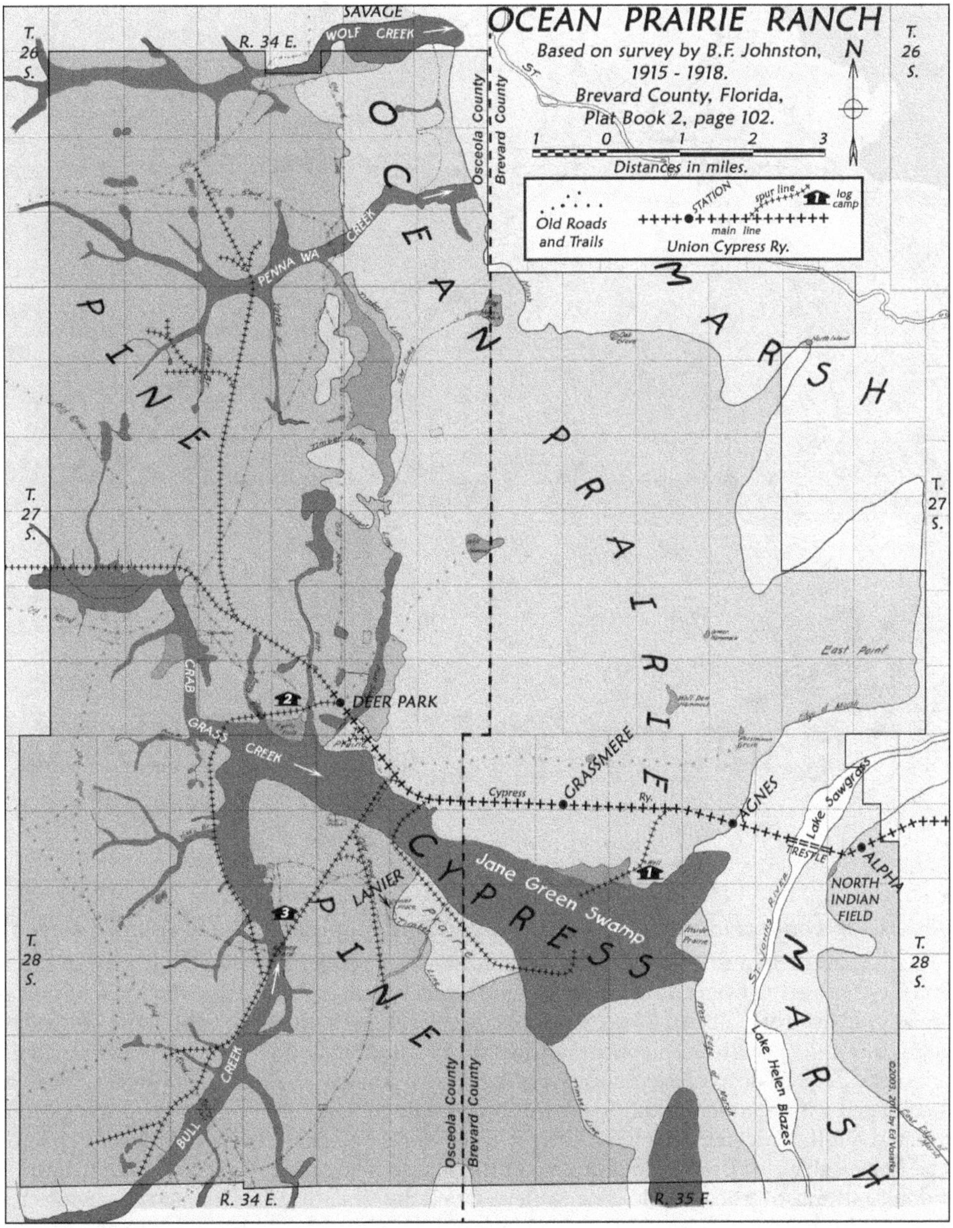

Logging operations needed to be expedited, so the railroad was built only as far as needed to reach the cypress. The first public excursion to Jane Green was scheduled for September 29, 1912. Crews started cutting and sending timber to the mill behind road engine No. 2 (arrived in May 1912), before the mill began sawing on October 28. Engine No. 1 (arrived in November 1911) was used to move cars at the mill. These were the only locomotives owned by the company until road engine No. 3 arrived in September 1913. The railway extended "down to the Platt place" (i.e., Deer Park) by late May 1913 and by February 1914 into the cypress near the confluence of Crab Grass and Bull Creeks (see map on page 60). The next areas to be logged were south along Bull Creek and west along Crab Grass. The Crab Grass spur followed the routes proposed for the CF&IR Railroad in 1904 and the planned Union Cypress line to Narcoossee (see map on page 46). (Author's illustration.)

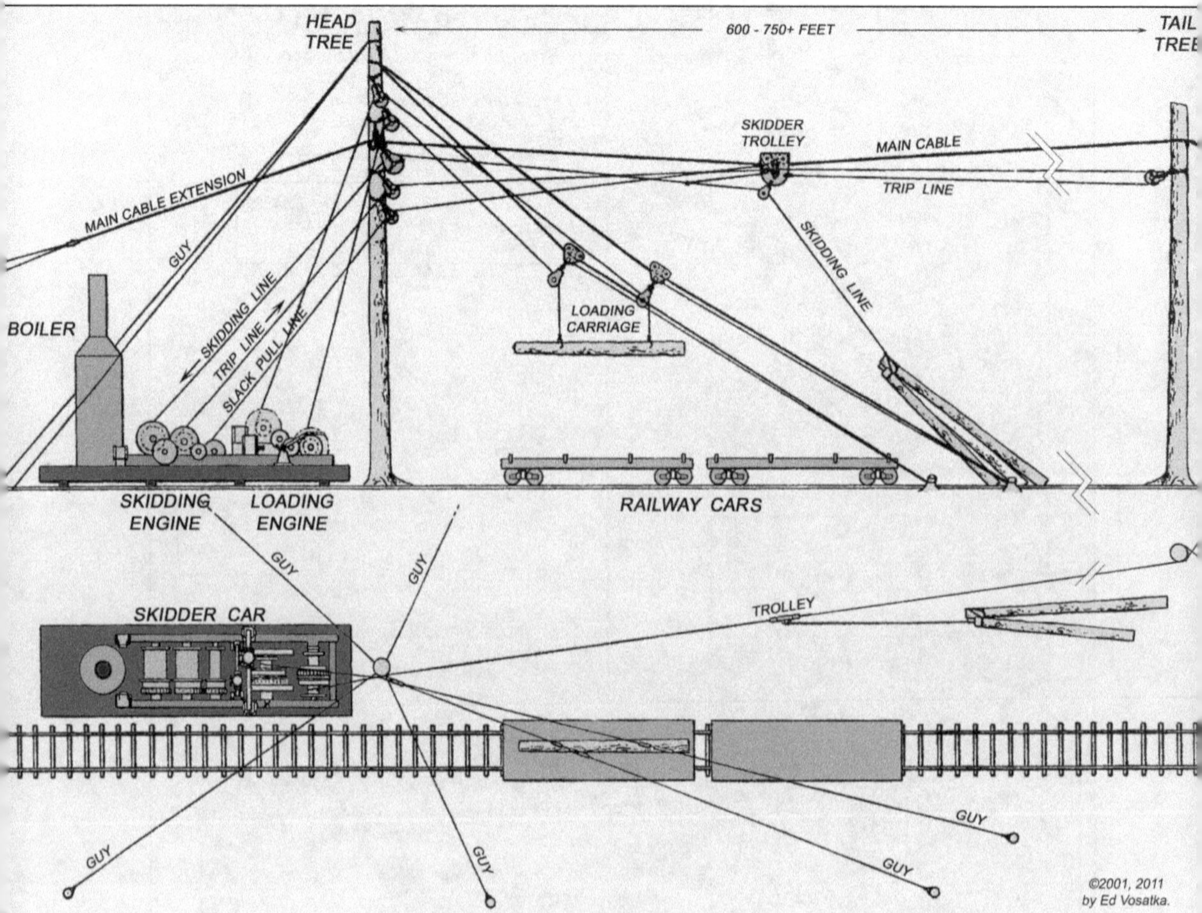

Lidgerwood overhead cable skidder and loader. Lidgerwood Manufacturing Company, New York City. Adapted from U.S. Patent No. 1,559,856, November 3, 1925.

Union Cypress used an overhead cable skidder and loader system built by Lidgerwood Manufacturing Company of New York City. In May 1913, it cost $2,425. Lidgerwood was probably the best-known developer of the specialized machinery that was required to log in difficult situations, from mountain valleys to deep coastal and riverine cypress swamps. A skidder's all-steel railway car was equipped with a boiler, skidding engine, and loading engine. Special railroad trucks could be rotated a full 90 degrees, allowing the car to be moved sideways onto a platform beside the track. Three cables led from geared drums on the skidding engine. A "trip line" pulled a trolley out along a strong, overhead cable. The trolley carried the "skidding line" out to the sawing crew, who secured it to the ends of the logs. A "slack pull line" controlled the slack in the skidding line. The skidding line then pulled the trolley back to the skidder, lifting and dragging the logs along. Logs were dropped at a "landing," where they remained until loaded onto railway cars by the loader, a separate engine and cable system. (Author's collection.)

Cableway skidding and loading practices were virtually the same throughout the cypress belt. In 1910 Louisiana, a typical crew and their daily wages, exclusive of board, included the skidder crew: skidder leverman (operated skidder, $3.50); fireman (operated boiler, $1.50–$2); tong hooker (put skidding line on log, $2.25); one or two hooking helpers ($1.75 each); signal man (relayed orders from foreman to leverman, $1.50); head rigger (attached rigging to trees, $3.50); two rigging helpers ($1.75 each); run cutter (cleared path for log to slide to landing, $2); and tong unhooker (removed skidding line from log at landing, $1.50); and the loader crew: loader leverman (operated loader, $2.75); ground loader (attached loading hooks to log at landing, $1.75–$2); and top loader (stood on top of load, placed logs, and removed loading hooks from log, $2.25). (Both, author's collection.)

Deep in the Jane Green, at least eight men are pictured standing in water while working on logs pulled in by the skidder, sitting off to the left. Beyond the men are two previously logged runs. Overhead cableway skidding would damage standing timber, but it was the only power system that did not appreciably destroy seedlings. (Author's collection.)

Two more logs are arriving at the landing, their ends held aloft by cables from the overhead trolley, which is just visible at the top of the photograph. Log cars are already in place to be loaded. A man standing on the car nearest the skidder gives a clue to the size of the equipment. (Author's collection.)

The Union Cypress advertised "red cypress" (see pages 39, 66, and 96), although the *Southern Cypress* (USDA Bulletin 272, September 27, 1915) described cypress from deep St. Johns swamps as being amber or light orange-brown, or "yellow" by trade name. It was once widely believed that cypress with dark-colored heartwood was more durable. A 1920 US Forest Products Laboratory study, though, showed that color made very little difference. Durability is entirely due to the amount of sapwood, which will decay. Size for size, trees with dark heartwood have less sapwood to rot away than trees with light heartwood; therefore, they last longer. The photographs on these two pages are from January 1915. (Both, author's collection.)

The Union Cypress provided excursions for many groups. Cadets from the Kentucky Military Institute, winter-quartered near Eau Gallie, were on this trip, probably taken on April 23, 1917. Based in Lyndon, Kentucky, the KMI started spending their winters here in 1906 and continued to do so until 1923 when they moved to Venice, Florida. The photograph below shows details of the patented "log dogs" (US patent No. 712,931), used in lieu of chains for keeping the logs on the cars. The photograph at the top of page 82 shows them in use. (Both, Ann Flotte.)

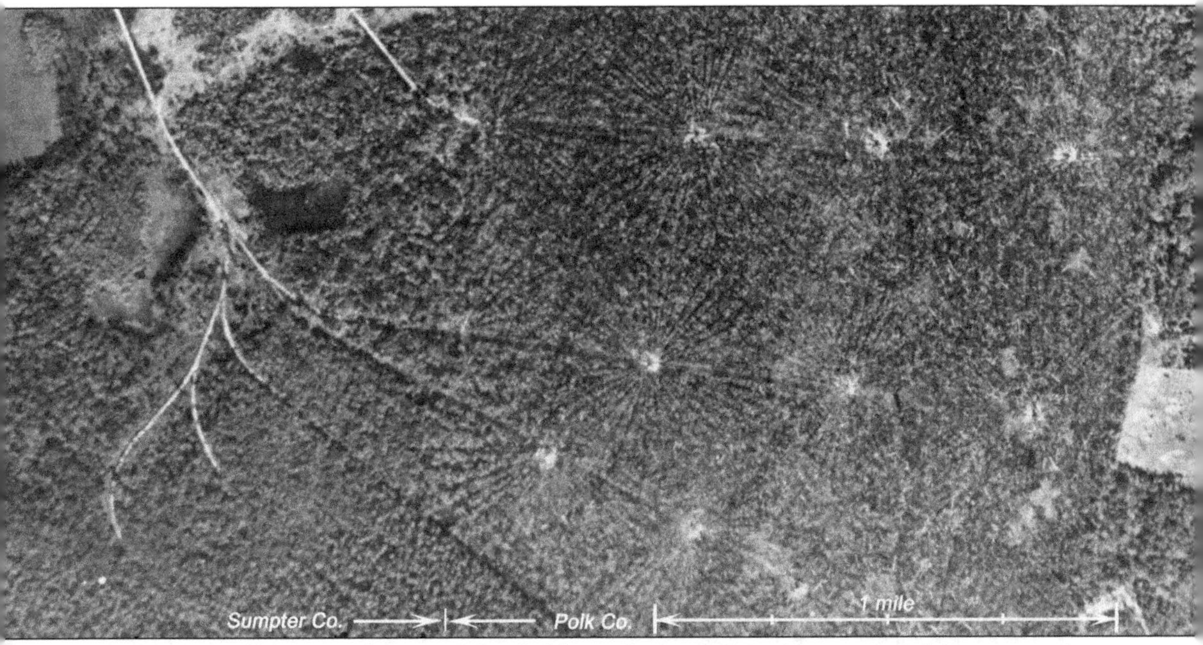

Logging proceeded in a circle around the skidder as the overhead cable was passed from one tail tree to the next, producing a starburst pattern to the cut. A daily cut averaged 35,000 to 45,000 feet. When logging was finished at one "set," crews picked up and moved a few thousand feet down the spur, repeating this operation over and over until the entire area was cleared. Once a spur had been completely logged, workers removed everything and started a new spur, even reusing the rail from the old spur. The method of clean-cutting by overhead cableway skidding resulted in almost complete destruction, practically eliminating the possibility of a future cypress stand. Even though cypress will successfully sprout from the stump (except when girdled) it typically takes 350 to 400 years to reach a diameter of three feet and perhaps 800 to 1,000 years to reach six or seven feet. This USDA aerial (flight negative CTS-3B-53) shows logging scars in the Withlacoochee Swamp on February 13, 1941. (Author's collection.)

A giant cypress fell in January 1917, southwest of Deer Park, after being girdled and deadened over the past 6 to 12 months. Felling was done by hand using two six-foot crosscut saws welded end-to-end to get the length needed. Company officials pictured in the center are secretary C.C. Burgess (sitting) and paymaster Bob Rowe (standing). One of the others may be the saw filer Arthur Plowden. (Bob Rowe.)

This was called a "four-log tree" because the "bole" (trunk without limbs) would make four standard logs. Each was cut somewhat over 16 feet to allow for trimming at the mill. This bole contained 19,600 board feet: butt log 6,500 feet, second log 5,000, third log 4,600, and fourth log 3,500. Per contract, Hopkins received $78.40 for just the trunk of this tree. The amount of wood in the limbs in unknown. (Bob Rowe.)

Smooth tracks were not guaranteed—not even when laid on cribwork. On these tracks, a skillful engineer with a small, light locomotive moved at a crawl to shunt a handful of cars at one time. Once logs were loaded, he moved the cars to another location where they were assembled into a train and then taken to the mill by a big road engine. Here, sitting on a log is C.C. Burgess. (Bob Rowe.)

C.C. Burgess (right), clerk Richard Henley (top log), and two unidentified men are dwarfed in this view of the back of the log pile. The straight, bare tree in the background is the headspar tree standing next to the skidder, which is hidden behind the top-most log. (Bob Rowe.)

A single 20-car train brought the entire tree to the mill, averaging over 4,000 feet of lumber per car. Union Cypress owned 45 of these 35-foot all-steel log cars, built by Magor Car Company of Passaic, New Jersey. Logging cars had old-style link and pin couplers because their flexibility allowed them to remain coupled over uneven twisting track, whereas knuckle couplers will disengage. These couplers were banned from interchange service in 1893. (Bob Rowe.)

This was not the company's only giant. Another, eight feet in diameter, resembled a vertical log, standing straight up from the ground, 120 feet to its first limb. It was cut in April 1919, yielding 6,000 feet of lumber. Going to Odessa, near Tampa, in March 1917, the Union Cypress cut a 10-foot-diameter giant, producing over 22,000 feet of lumber. (Bob Rowe.)

The largest native Florida tree is the "Senator," once part of a group of four used as landmarks by Native Americans and pioneer settlers. Over 3,500 years old, it is 17.5 feet in diameter and contains approximately 50,000 board feet of lumber. Reported to have been 165 feet tall until damaged by a hurricane in 1925, it now stands at 118 feet with an average crown spread of 57 feet. It is registered as the Florida Champion Baldcypress under the American Champion Tree Program, created by the American Forests organization in 1940. American Forests publishes their *National Register of Big Trees* every two years. The Senator is in Big Tree Park, just off US 17-92 north of Longwood. The park, now administered by Seminole County Parks and Recreation, was dedicated by former president Calvin Coolidge in 1929. (Author's collection.)

THE KAISER

THE BEAST of BERLIN

The Great War Picture That Has Created Such a Wonderful Interest Throughout The Country.

At the Crescent Theatre, Hopkins

WEDNESDAY, AUGUST 14th

Two Performances, 7 and 9 p. m.

50 cents--One Price to All

During the great war, the Crescent Theatre showed patriotic movies in addition to regular silent films, including *The Spy*, *Over There*, *The Kaiser: The Beast of Berlin*, and *The Fourth of July in Paris*. Herbert "Food Will Win the War" Hoover requested that women register for membership with the US Food Administration and take the pledge "to carry out the directions and advice of the food administrator," reported the *Cocoa Tribune* on July 19, 1917. People "Hooverized" with wheatless, meatless meals: each week, 11 meals without wheat products, nine without meat products, and three without pork products. During "Potato week" in April 1918, everyone was expected to eat potatoes three times a day without any wheat products on the side. On April 24, 1918, the *Melbourne Times* implored residents to "eat potatoes and send the wheat to our boys 'over there.' " People planted their own Victory Gardens, saved fruit pits and shells for gas mask filters, and enjoyed daylight savings time. This show listing appeared in the *Melbourne Times* on August 7 and 14, 1918. (*Melbourne Times.*)

Five

Conflagrations

Far to the east, across the Great Pond, a terrible storm was brewing, with thunder and lightning, the smell of gunpowder, and the stench of death. President Wilson declared war on April 6, 1917, and the first civilian registration for military service followed on June 5. "Going to Cuba or Canada Will Not Save Slackers," proclaimed a *Melbourne Times* headline on May 30, 1917. In addition to a man's name, birth date, occupation, and so on, registrars also recorded race: Melbourne—31 "white" and 14 "colored" and Hopkins—six "white" and 65 "colored." Leon H. Martin, the Union Cypress doctor, volunteered for the Medical Reserve Corps in early June. The first Yanks arrived in Paris on July 4, 1917.

Further registrations were held on June 5, August 24, and September 12, 1918. Over the course of the war, 15 Melbourne men entered the Army and four entered the Navy. From Hopkins, 29 joined the Army and one joined the Marine Corps. Twenty Deer Park men joined the Army.

President Wilson created the US Railroad Administration (USRA) on December 26, 1917, placing the nation's railroads under government operation and control because unusually high rail traffic for carrying on the war was causing congestion and delays nationwide. Wartime meant government orders took precedence, making it difficult for railroads to secure lumber for their needs. Hopkins shipped 50,148,229 pounds of freight, mostly lumber, over the Florida East Coast Railway from November 1916 through October 1917, far exceeding any other location in Brevard. In 1919, they shipped 56,506,362 pounds, again the highest in the county. Governmental control over the railroads was finally lifted on March 1, 1920.

By May 1918, Union Cypress employment had dropped to 236. Then in the fall, Spanish influenza started slowly slipping in. By mid-October, the FEC was practically shut down from the lack of engineers. Day after day, the Hopkins Red Cross ladies carried hot soup to the sick in Hopkins and the lumber camp at Deer Park. Only one person in south Melbourne died from the flu.

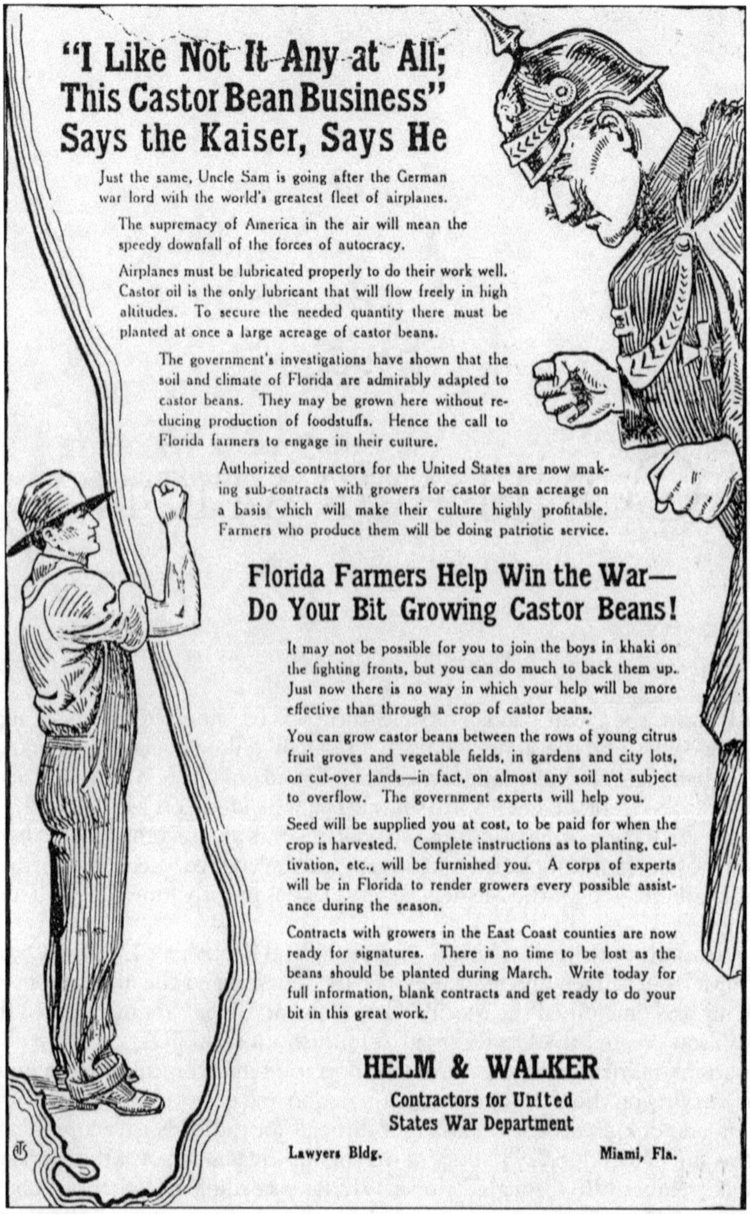

Since the early 1900s, castor oil has been the preferred lubricant for the rotary aeroplane engines developed in Europe. During the war, it was the oil most universally used by Allied aircraft. Demand was so great by early 1918 that the US government started contracting with local industries to supply castor beans. A contract would be issued for 10,000 acres or less, and the contractor was required to furnish a good faith bond of $1 per acre. The government furnished the seeds at cost plus transportation and paid $3.50 per 46-pound bushel for all of the beans delivered to the nearest railway station. Their subcontractors (farmers) were paid $3 per bushel. Castor beans grew well in this area. This advertisement appeared in the *Melbourne Times* on February 20, 1918, and on June 5, 1918, the newspaper reported that "soon the children's favorite will be flowing, and so far as known, the children are willing for the government to have it all—for the air planes." (*Melbourne Times*.)

Four nationwide Liberty Loan Bond drives were held to raise money for the war effort. During the first two, in 1917, Melbourne and vicinity raised $115 and $9,650. During the third in 1918, they raised $15,700, for which Melbourne was awarded a Liberty Loan Honor Flag. During the fourth in late 1918, they raised over $25,000. This illustration appeared in the *Melbourne Times* on October 9, 1918. After the armistice, a Victory Liberty Loan bond drive was held in April and May 1919. (*Melbourne Times.*)

The Brevard County Red Cross Chapter, organized in Cocoa on June 6, 1917, was the third chapter in Florida, the others being in Jacksonville and Miami. South Melbourne and Hopkins formed the Colored Auxiliary to the Melbourne Red Cross Branch on May 6, 1918. During the second Red Cross fund drive in May, Hopkins topped its $400 assessment by raising $642, mostly by the 236 Union Cypress employees. (*Melbourne Times.*)

During the war, daily lumber production increased to 60,000 feet. The mill was operating around the clock filling rush orders when, around 4:00 a.m. on June 9, 1917, the planing mill and three railway cars were completely engulfed by fire, an uninsured loss of about $30,000. A new mill was built and in operation by mid-August. Just after midnight on the 25th, it also caught fire, leading the *Melbourne Times* to state on the 31st that "it certainly looks as if someone was trying to do away with it as well as the immense amount of lumber stowed in their yards." In late October 1918, engine No. 3 was badly damaged when she broke through the Platt Branch trestle west of Deer Park. The locomotive could not be moved and was blocking the main line, so the mill had to be closed until the engine could be repaired and moved to clear the line. Speculation was that the light rails had spread and that the locomotive dropped between them and through the trestle. (Bob Rowe.)

BREVARD COUNTY'S

Honor Roll of Heroes

These Men GAVE ALL

The Government Asks YOU Only to Lend

FRANK B. HUDDLESON	Melbourne
ELI LIGHT, Jr.	Melbourne
JOHN McMILLAN	Eau Gallie

MELBOURNE CELEBRATES---TWICE

Took No Chances on First News of Peace Being a Hoax---But Big Noise Took Place Monday

Melbourne fell for the fake news dispatch to the effect that the war was ended last Thursday, just as did all of the other important citie[illegible] [illegible] countr[illegible]

Of the 62 men who entered the service from Melbourne, Hopkins, and Deer Park, 33 went overseas. Sgt. Harold Ward, 19, was wounded and Pvt. Frank Huddleston, 24; Cpl. Rufus Ridgdill, 28; and Pvt. Eli Light Jr., 28 (all from Melbourne) were killed in action. Pvt. Fletcher McWhite, 28, of Deer Park, was murdered overseas a month after the armistice. John P. McMillan, 21, died of disease in Texas during training. Sgt. Joseph M. Hopper, 25, of Hopkins, was awarded the MCC. Dr. Leon Martin, 30, former Union Cypress company doctor, was awarded a Silver Star Citation for gallantry on the front at Nantillois, France. His story ran in the *Melbourne Times* on November 20, 1918: "Captain Leon Herbert Martin . . . Captain Frank Wheeler . . . Lieut. Samuel Marks . . . on Sept. 28th volunteered to go up to the front line dressing stations and all did wonderful work carrying stretchers and giving first aid while under shell fire. For three days these men, although not asked to leave their station in the rear, rendered medical attention and carried water and food to the wounded." These clippings ran in 1918 in the *Melbourne Times*, on October 2 (above) and November 13 (right). (Both, *Melbourne Times*.)

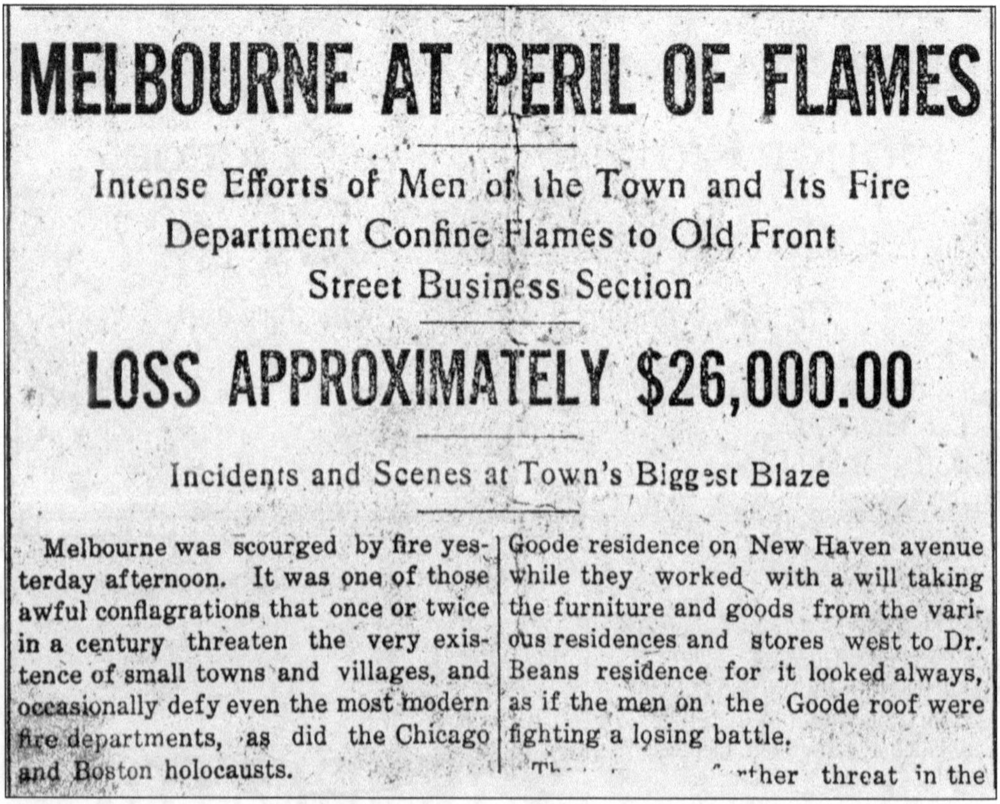

MELBOURNE AT PERIL OF FLAMES

Intense Efforts of Men of the Town and Its Fire Department Confine Flames to Old Front Street Business Section

LOSS APPROXIMATELY $26,000.00

Incidents and Scenes at Town's Biggest Blaze

Melbourne was scourged by fire yesterday afternoon. It was one of those awful conflagrations that once or twice in a century threaten the very existence of small towns and villages, and occasionally defy even the most modern fire departments, as did the Chicago and Boston holocausts.

Goode residence on New Haven avenue while they worked with a will taking the furniture and goods from the various residences and stores west to Dr. Beans residence for it looked always, as if the men on the Goode roof were fighting a losing battle.

Th[illegible] [illegible]ther threat in the

Tuesday, February 18, 1919, was a clear, 66-degree day, with a 40 mile-per-hour northeasterly wind. It had been dry, since only 0.22 inches of rain had fallen over the previous 12 days. January was 1.06 inches below normal. At about 4:00 p.m., upstairs in the Goode building beside the city dock, someone took a shot at a lit kerosene heater, then threw it out the back window. It burst onto the wood boardwalk, setting it afire. Shooting flames 100 feet into the air, the spreading fire took less than an hour to consume all Front Street buildings and the foot of the city pier. Citizens formed a bucket brigade and saved a residence at the corner of New Haven Avenue, checking the fire's westward advance through residential areas toward the rest of the town. The *Melbourne Times* carried this story the following day. (Above, *Melbourne Times*; below, Weona Cleveland.)

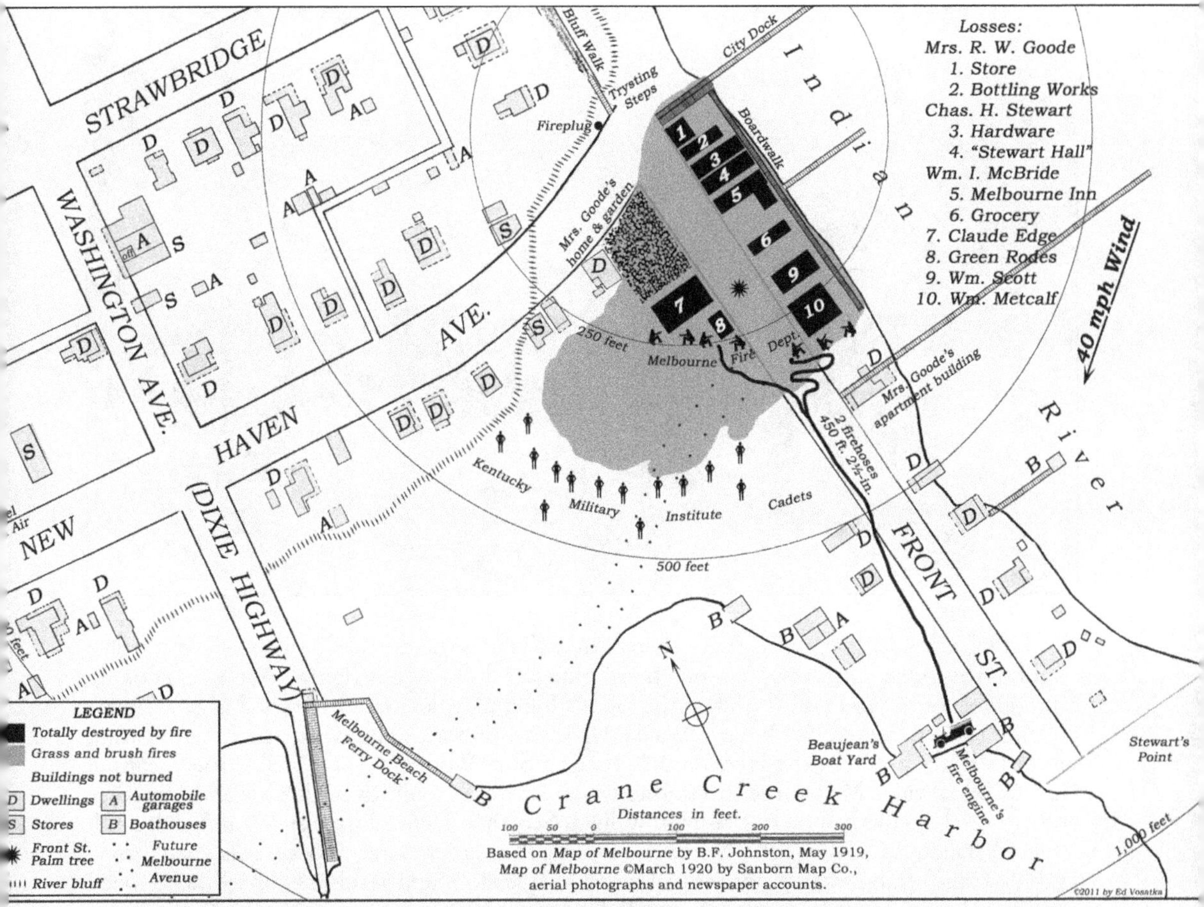

DOWNTOWN MELBOURNE FIRE, FEBRUARY 18, 1919.

Melbourne's 18-man volunteer fire department arrived with its 1916 Waterous fire engine, but flames and intense heat thwarted their attempt to reach the fireplug on Front Street. They headed south, tearing through Mrs. Goode's garden and other backyards, cutting fences en route to the lower harbor. As they went, they played out two 450-foot lengths of 2.5-inch fire hose. Pumping water from the river, firemen stopped the fire's southward advance before it reached Mrs. Goode's apartment building. Cadets from the Kentucky Military Institute near Eau Gallie battled the large grass, brush, and palmetto fires spreading southwestward. Eau Gallie's volunteer fire department responded, but their dilapidated gasoline engine pumper, which had to be towed, careened off the Eau Gallie River bridge and into the river. By 6:00 p.m., it was all over. The February 19th issue of the *Melbourne Times* reported, "Had there been a building on the southwest corner of Front street and New Haven avenue, (instead of Mrs. Goode's garden) there would have been no stopping the fire east of the Dixie Highway or north of the harbor." (Author's illustration.)

The *East Coast Advocate*, on February 28, reported, "The fire made a clean sweep of every article on all of this property in less than one hour's time." All of these buildings were lost. Looking south along Front Street on the left is the Goode building where the fire started, then Goode's bottling factory (rented by Florence Wackerly), Chas. Stewart's Hardware and funeral parlor, Stewart's Hall (rented by the Free Masons, Eastern Star, Woodmen, and Melbourne Cleaning Co.), C.T. McBride's Melbourne Inn (formerly C.J. Hector's Hotel Riverside), McBride's grocery and dry goods (only the porch is visible), William Scott's residence (not visible), and William I. Metcalf's building. Crossing Front Street and coming this way are Green Rodes's residence (barely visible), Claude Edge's two-story general merchandise store and residence, and Mrs. Goode's garden. (Above, Weona Cleveland; below, Bob Gross.)

Looking north along Front Street toward the bluff and Trysting Steps, the remains of Claude Edge's store and two single-cylinder gasoline engines appear to be the only things left standing. Even the big palm tree in the middle of the street is devastated. Losses included Claude Edge, $12,000; Mrs. Richard W. Goode, $3,000; C.H. Stewart, $3,000; Green Rodes, $800; William I. Metcalf, $1,200; William M. Scott, $700; and C.T. McBride, $6,000. Only Claude Edge was insured, for $3,800. With the local economy no longer depending on river trade, the old downtown was abandoned and a new downtown was built around the railway station. (Above, Weona Cleveland; below, Bob Gross.)

Sunday, August 31, 1919, was clear and about 94 degrees. The sawmill was shut down for the Sabbath. Around 4:00 p.m., the watchman discovered a small blaze in the northwest corner of the mill, perhaps a spark from the trash burner. Before he could get a pail of water, the fire burst forth in the sawdust and other inflammable material and spread rapidly. The mill's sprinkler system activated, but the fire soon reached the power plant and destroyed the boilers, engines, and fire pumps. Water from the 30,000-gallon elevated tank kept the sprinklers going until all the water was gone. Melbourne's fire department pumped water from the log pond—two streams uninterrupted for nine hours. The Eau Gallie department arrived within an hour. Flames were under control in about two hours, but piles of wood and sawdust smoldered until after midnight. The big mill was a $100,000 loss, with only about a third being insured, but much of the plant and about five million feet of lumber in yards were saved. Photographs of the mill fire have not been found; this is the McGowin-Foshee mill fire in the 1920s. (Florida State Archives, photograph n032410.)

SOUTHERN LUMBERMAN

DECEMBER 13, 1919

For Sale—Second-Hand Machinery

FOR SALE

Equipment of the Union Cypress Company, Hopkins, Fla., in Liquidation. 1 Lidgerwood Cableway Skidder, late model, complete. 3 Locomotives. 30 All-Steel Cars. 30,000 Gal. Steel Tank, 75 Ft. Tower, 1 Mile of Water Mains, Hydrants and Hose. 5 Steel Inclosed 150 H. P. Return Tubular Boilers. 2 Engines. Machine Tools. Machinery. 5 Miles 12-lb. Yard Rail, and Lumber Carts. $20,000.00 New Stock in Storehouse. Also 20-Ton Browning 8-Wheel Locomotive Crane. 20 M Capacity Circular Sawmill, complete, and 3 Miles 35-lb. Relaying Steel Rails and 16-Ton Forney Type Locomotive. 1 Yard Marion Floating Dipper Dredge. 2 Large Worthington Compound Fire Pumps, and 2 Cameron Compound Pumps. 2 300-K. W., 440-Volt D. C. Generators; 1 200-K. W., 440-Volt D. C. Generator; Switchboards and Motors. 1 60-Ton, 10-Wheel Locomotive, newly overhauled. 1 45-Ton Baldwin Mogul, just out of the shop. Also 1 Complete 9-ft. Right-Hand Band Mill. 1 Complete 8-ft. Double Band Mill. 1 50 M Capacity Circular Sawmill, complete. Address

H. L. CHURCH.

P. O. Box 209, 212 West Forsyth St., Jacksonville, Fla.

Compared to wood, a steel and concrete building is more expensive to build, repair, alter, and expand, and it is noisier. Its permanence and, most importantly, its inherently lower fire hazard far outweigh its disadvantages. And, it is lightning-proof. However, if not kept clean, it can be seriously damaged by a fire. Unprotected steel framing is, by itself, not fireproof. Metal columns and trusses quickly fail by bending at high temperatures (wooden ones are consumed slowly). The chief reason for not fireproofing is the extra initial, and often prohibitive, cost. How much equipment was sold at this liquidation sale and who bought it is not known. Of the locomotives, the "Forney-type" was No. 1 and the 10-wheeler (4-6-0) could have been Baldwin No. 4. The "Baldwin Mogul" (2-6-0) could have been either No. 2 or No. 3, neither of which were Baldwins. No. 3 was sold in mid-January 1920, and the fate of No. 2 is unknown. Shipments from Hopkins during 1920 were one-half of what they had been the year before the fire (see page 85). (Anonymous.)

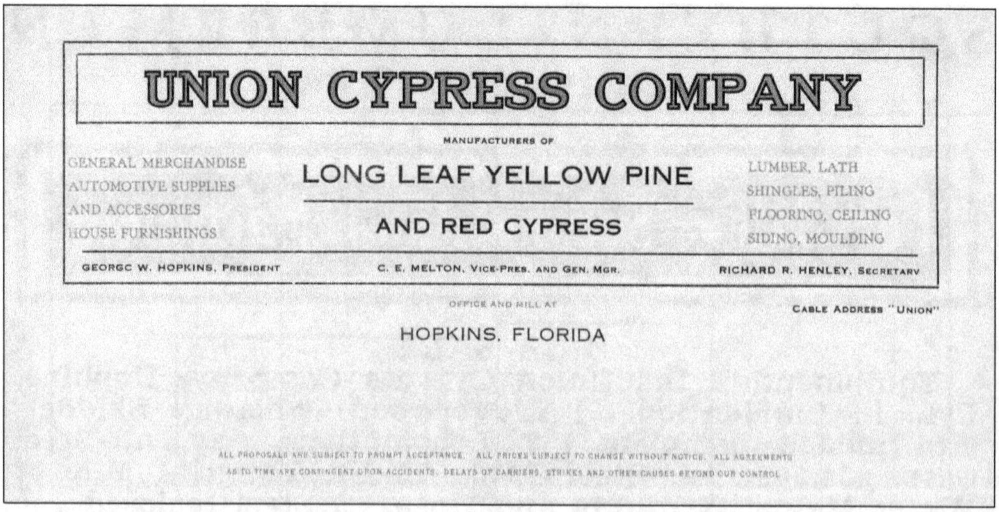
UNION CYPRESS COMPANY

MANUFACTURERS OF

LONG LEAF YELLOW PINE

AND RED CYPRESS

GENERAL MERCHANDISE
AUTOMOTIVE SUPPLIES
AND ACCESSORIES
HOUSE FURNISHINGS

LUMBER, LATH
SHINGLES, PILING
FLOORING, CEILING
SIDING, MOULDING

GEORGE W. HOPKINS, PRESIDENT — C. E. MELTON, VICE-PRES. AND GEN. MGR. — RICHARD R. HENLEY, SECRETARY

OFFICE AND MILL AT

HOPKINS, FLORIDA

CABLE ADDRESS "UNION"

ALL PROPOSALS ARE SUBJECT TO PROMPT ACCEPTANCE. ALL PRICES SUBJECT TO CHANGE WITHOUT NOTICE. ALL AGREEMENTS AS TO TIME ARE CONTINGENT UPON ACCIDENTS, DELAYS OF CARRIERS, STRIKES AND OTHER CAUSES BEYOND OUR CONTROL

By January 1, 1920, only 68 employees remained on the payroll. Many of the former workers had moved to Bridgend, on the St. Johns just north of Lake Harney, to work at the new Osceola Cypress Company sawmill. In August, Sanford banker Forrest Lake purchased all the Union Cypress interests, including cypress timber and railway. A new circular sawmill was erected southeast of the burned-out mill. Sawing began August 13, 1920, producing 40,000 feet daily of yellow pine lumber, plus shingles and lath. By early 1923, the company had 200 employees and a monthly payroll of $15,000. The *Melbourne Times* carried this advertisement from December 14, 1921, through January 19, 1925. The mill and railway ran until G.W. Hopkins's death in January 1925, when all operations ceased. (Above, author's collection; below, *Melbourne Times*.)

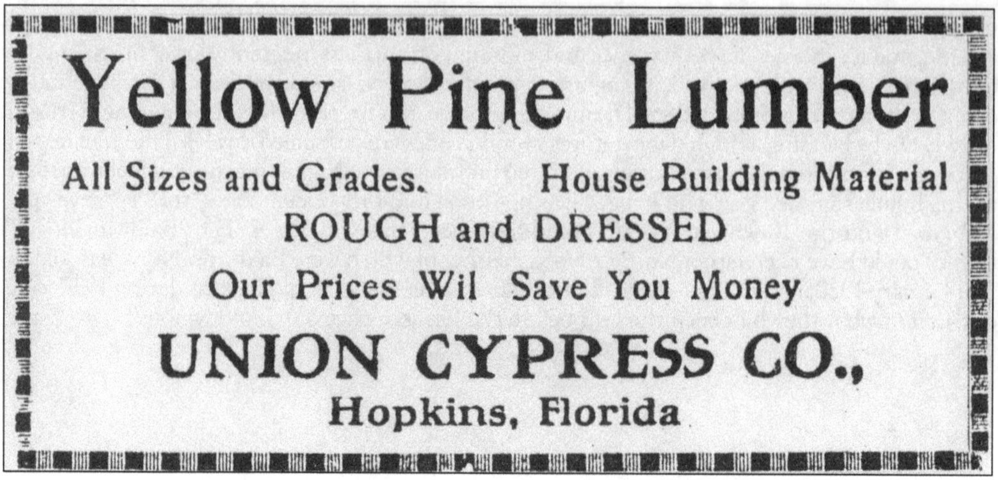

Six

On the Trail of the Lonesome Pine

After the mill burned, Melbourne and Hopkins were dark. Eau Gallie offered to supply power to Melbourne, a contract was made, and a line was constructed. Four months later, on December 31, the lights came back on, but only from sundown to daylight. Electricity was extended to South Melbourne in July 1920, and full 24-hour service to Melbourne was restored August 19. In October, Melbourne offered to supply power to the Union Cypress.

George Hopkins owned 240 million feet of pine (a bookshelf from Melbourne to Los Angeles and back). In 1919 and the early 1920s, he bought out all the minor stockholders and purchased $20,000 stock in the corporation. On March 19, 1920, C.E. Melton purchased all 3,148 shares then owned by Mr. and Mrs. Hopkins for full par value of $314,800, with annual payments of $50,000. Melton became vice president and general manager, and Richard Henley became secretary. An amendment to the original timber contract proposed to eliminate hardwoods from consideration and increase Hopkins's cypress and pine payments to $5.60 and $3 per 1,000 feet, respectively. However, another amendment on December 15, 1920, allowed hardwoods and gum to be cut and refined Hopkins's payments to $2 per thousand feet for all timber.

Only four distinctly Southern species of pine were actually cut for the market: long-leaf, Cuban, short-leaf, and loblolly. Pine, however, appeared on the market under a variety of local, market, and lumbermen's names, causing confusion among architects, builders, and engineers as well as lumber dealers and lumbermen themselves. Throughout the Southeast, the long-leaf species was known by 22 other common names, Cuban by 5, short-leaf by 11, and loblolly by 18. Moreover, several of these common names were applied to more than one species, and, to make matters worse, these names changed over time.

Yellow pine forests were the source of most of the lumber used in the South and were the chief source of railroad lumber used in the East and South. In February 1908, the American Railway Engineering and Maintenance of Way Association (AREA) standardized the names for structural timber used on railways. "Southern yellow pine" thence collectively included the long-leaf, short-leaf, and loblolly.

No matter how elaborate the equipment, it still came down to two men on a crosscut saw. In addition to the Hopkins contracts, other contracts were signed in 1922 to log pine southwest of Melbourne. For their 7,540 acres of pine, Frank C. Powell and Thomas J. Shave received $2.50 per thousand feet for the first 10 million feet cut (measuring eight inches or more) and $1.25 per thousand feet for all timber in excess of 10 million. James A. Lepper received $2 per thousand feet for the lease of a 20-foot right-of-way across his 640 acres near June Park, along with any timber on it. For small trees cut to make cross ties, he received 2¢ per tie. (Left, author's collection; below, author's illustration.)

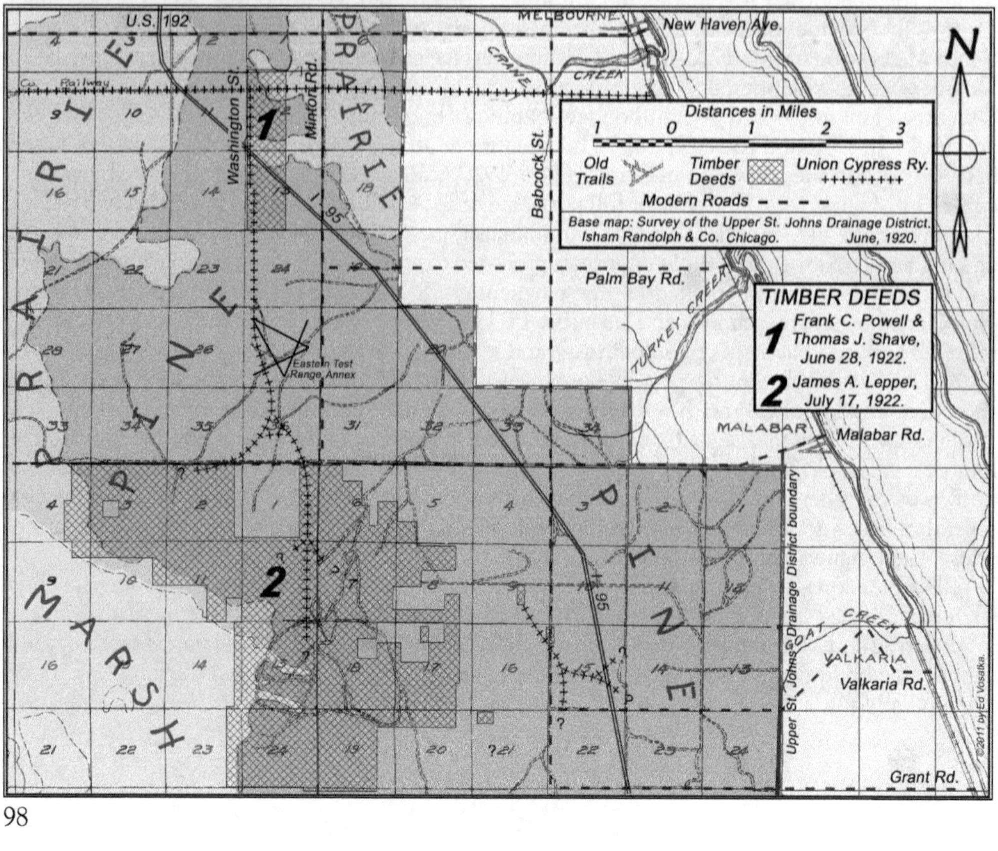

The Union Cypress owned a steam-powered ground skidder and a log-loader, like those shown in these undated catalog cuts, both built by Clyde Iron Works of Duluth, Minnesota. The self-propelled skidder (above) carried a boiler; either two, three, or four independent skidding units (each being an engine, geared drum, and 1,000 to 2,000 feet of skidding cable); and one or two heavy booms. A crew consisted of 19 men. Nine draft animals pulled the skidding lines out to the sawing crews. The McGiffert log loader (below) was patented in 1902 by John R. McGiffert. In 1914, McGifferts cost between $4,500 and $6,000. Although the loader could travel under its own power at seven to eight miles per hour, it was usually towed to the logging site. Six men made up the crew. (Both, CabincreekCDs.)

Union Cypress No. 5 was a used Glover 0-6-0 (c/n 101,413), purchased May 9, 1921, from the Yellow Pine Operating Company at Maytown, Florida. She was built in October 1912 for B.F. Brinson of Lake City, Florida, and Yellow Pine acquired her in August 1918. Her six driving wheels were the size of freight car wheels (31 inches) and formed a 5.5-foot rigid wheelbase. On the side of this obviously jury-rigged tender was an injector and suction hose, which the crew used to refill the tender with water from any natural supply along the route. Fifteen ex-M&O (Mobile & Ohio Railroad) gondolas were acquired for $5,344.80 in February 1924. They originally came from the East and West Railroad of Alabama. (Above, Cliff McNulty; below, Carolyn Kempfer.)

Seven

Tin Lizzie and the Iron Horse

That single-lane "road" across the St. Johns from 1917 was so poorly built that it soon became a target of jokes and criticism. In January 1921, the story reached the *Chicago Sunday Tribune,* which was paying $1 for each letter published describing the worst stretch of road the writer encountered. In May, an Osceola County Grand Jury presentment criticized the road situation.

In October, the State Road Department designated it a state aid road, allowing it to be built with federal and either state or county funds under SRD's supervision. Counties would pay at least half the cost, with the state paying the remainder. Building a road across two and a half miles of marsh, which in wet times might flood 10 miles wide, became a nightmare. Work fell behind schedule, contracts were not enforced and were allowed to lapse, and contractors defaulted. Everyone blamed the overseer. The *Kissimmee Valley Gazette* of October 7, 1921, commented editorially that "getting performances from promises out of the state road department is about as satisfactory as attempting to stuff butter in a wildcat's ear with a hot awl." So much soft muck had to be dredged out to reach sand and clay suitable to support the road fill that a contractor told the *Cocoa Tribune* on July 20, 1922, "the other side of the earth seems to be just below where we are working."

Brevard County closed its part of the road to traffic in early March 1923. When the river flooded in December 1923, crews worked in 12 feet of water using a suction dredge to build the 30-foot fill.

"If George Washington had not worked up more speed when crossing the Delaware than our State Road Department shows in crossing the St. Johns, me an' you, Nancy, would be British subjects today," opined the *Eau Gallie Record* on August 28, 1924. Finally, in July 1924, the Melbourne Chamber of Commerce turned to Governor Hardee for help, asking him to investigate. As a result, the SRD announced in August it would do the work itself rather than rely on outside contractors. The highway reopened January 21, 1925, three years after the first contract was let for work to be completed within 120 days.

As a courtesy, the Union Cypress in March 1923 began carrying automobiles across the St. Johns valley on flatcars. Wooden ramps, about six miles apart at Deer Park (left, in winter 1923) and Simon Road (pages 103and 104), allowed automobiles to be driven onto and off the flatcars (see map on pages 48–49). The Deer Park ramp was later moved eastward to Agnes Station (below and page 100). Up to 12 automobiles could be carried each trip for $1 per car, and passengers, and trips were scheduled every 30 minutes. The ferry operated at a loss the first season. Prompted by the Melbourne Chamber of Commerce and businessmen of St. Cloud, Kissimmee, and Melbourne, the Union Cypress resumed ferry operations September 17. Fares also increased to $2 for small cars and $2.50 for larger ones. (Left, Myrtle Svedelius; below, author's collection.)

Locomotive No. 6, another SI&E rebuild (c/n 1785), was purchased on October 6, 1923, for $1,250, to operate the ferry train. She was a Baldwin 4-4-0, built in 1873 for the Cincinnati Southern Railway. Subsequent owners were the New Orleans & North East Railroad; Vicksburg, Shreveport & Pacific Railway; and Victory Gravel Company in Ouchita Parish, Louisiana. By March 1924, the company had leased four 40-foot FEC Railway flatcars, seen here at the Simon Road ramp. Now the ferry train could handle 16 automobiles per trip and five round-trips were scheduled every day. It ran over eight miles of track and had become somewhat profitable, although earlier losses had not yet been overcome. Unloading 16 automobiles and loading another 16 took about one hour. The man with his back to the camera (above) is Vertner Snow, engineer, and the barefoot boy on the right is his son Francis. (Above, Vertner Snow; below, Cliff McNulty.)

The 6-spot waits at the Simon Road ramp as automobiles are off-loaded on December 16, 1923. Between the tender and the first flatcar is a long wooden pole with a steel link-and-pin connector at each end. This flexible arrangement allowed the train to remain coupled as the cars were switched onto a short spur track leading to the ramp, as in the photograph below. The photograph at left shows the pole attached to the tender's coupler. The knuckle has been removed from the coupler and is sitting on the deck between the feet of, from left to right, Francis Snow, Mrs. Swope, and Martha Snow. The pole is held in place by a pin inserted through it, and the knuckle pin holds in the coupler head. (Left, Francis Snow; below, Drew Huy.)

A six-day storm in October 1924 dumped 11.88 inches of rain, four inches of that falling on October 10 between 6:00 a.m. and noon. The ferry train became trapped in the St. Johns when water reached two feet over the trestle and flooded the engine's firebox. Marion L. Lassiter and his section crew are pictured here surveying damages from their gasoline-powered "puttcar" (US patent 1,469,741). Service resumed October 30 with two daily round-trips, and with four trips beginning November 3. In Hopkins, the FEC embankment became a dam, and 10 feet of water flooded the town until the embankment itself collapsed. Water rose in the commissary and washed over track at the railroad depot. Cocoa was without train service nearly 36 hours, and the Indian River and St. Johns waters merged in several places along the Dixie Highway. (Both, Sallie Madden.)

As steamboats gave way to trains, Melbourne's business district shifted west toward the railway. Building began in earnest following the downtown fire of 1919. Looking east (above) along Melbourne and New Haven Avenues are C.T. McBride's new store and the US Post Office. The FEC track is immediately beyond the post office. Looking west (below) along New Haven Avenue. Along the north side of the 75-foot right-of-way are J.P. Cooper's Drug store and a barbershop, Melbourne Furniture, Gem Theatre, Davis Department Store, Florida Indian River Land Company, Geiger Brothers Garage, Methodist Episcopal Church, Orange Avenue, and a three-story apartment building. Along the south are the bandstand and the Melbourne Hotel. The railway station and Brown House are out of view to the right. (Both, Weona Cleveland.)

Eight

From Boom to Bust, then Depression

Florida's great 1920s land boom set national records for numbers and speed, with over 300,000 new settlers in just two years, more than in California during the Gold Rush. Censuses recorded 968,470 Florida residents in 1920 and 1,253,600 in 1925. Melbourne had 533 residents in 1920, 1,245 in 1925, and 3,500 after the city annexed Hopkins on May 26, 1925. Melbourne was now the largest city in Brevard County.

Melbourne started 1925 with a New Year's concert at Stewart's Hardware on the evening of January 1. The concert, heard over station WEAF, New York City, was probably the first publicly broadcast concert heard in Melbourne. The Kissimmee Highway reopened January 21, led by a car from Maine. The city got sidewalks in June, a municipal sewerage system in July, and a municipal water system in December.

In January 1925, over 25 trains were passing through Melbourne every day. By August, the FEC had become overwhelmed transporting construction material and declared an embargo on all freight except fuel, petroleum, livestock, and perishables. To better feed the boom in South Florida, FEC doubled its tracks in early 1926. In December, they built a new combined Melbourne-Hopkins freight terminal on the sites of the Union Cypress Company's old office and commissary. Besides providing more storage and better freight services than the original Melbourne terminal, it gave downtown Melbourne relief from the incessant congestion and noise. The Florida real estate bubble burst during the summer of 1926 and finally died in September with the great Miami hurricane. But people keep coming. So many were killed at railroad crossings that on October 1, 1927, Florida enacted a state law requiring drivers to "stop, look, and listen."

The year 1926 also brought modern phone service on February 21 and home mail delivery on August 16. On October 13, Melbourne enacted an ordinance making parents responsible for children under 14 driving cars.

In January 1928, water lines were extended to Hopkins. In February, the old Crenshaw Hotel, on the corner of Potter Street and the Dixie Highway, became the new Melbourne Hospital. Melbourne's airport was designated a refueling station for the New York–Miami Air Mail route, with the first planes arriving on November 30. Air mail postage was 5¢ for the first ounce and 10¢ for each additional ounce.

Two of the four dredges used to dig the canals of the Melbourne-Tillman Drainage District were transported to their starting points by the Union Cypress Railway. American Steel Dredge No. 50 and a Quarter (dormitory) Boat were erected and launched at the railway 6.7-mile point (see map on pages 48–49). Starting November 12, 1923, it first dug a launching basin for Vulcan Iron Works Dredge No. 40, then began cutting its way west to start digging and building the district's outside levee (see map on page 50). This dredge was a revolutionary design from the American Steel Dredge Co. of Fort Wayne, Indiana. It could easily be dismantled, transported, and reassembled. Even the hull was built from bolted-together steel sections. A complete 20-by-80-foot hull, six feet deep, could be loaded on one flatcar. These images appeared in the 1924 *Melbourne Yearbook*. (Both, Ann Flotte.)

In July 1924, the Union Cypress began erecting a new, $150,000 pine and cypress mill on the site of the burned mill. The *Melbourne Times*, on August 6, 1924, reported the new mill was "constructed throughout with steel and concrete and will be as near fireproof as it is possible to build a structure of this character." Structural materials came from the former Rodman Lumber Company sawmill, pictured here, in Rodman, Florida, 10 miles southwest of Palatka. Rodman was a circular sawmill with a capacity of 100,000 feet per day. Products included cypress, long-leaf yellow pine, railroad ties, headings, shingles, staves, and lath. Its planing mill had a capacity of 30,000 feet. The company went bankrupt following the death of its president, Henry S. Cummings, in April 1922, and its mill was dismantled and sold at auction in May 1923. (Both, Jana Hyten.)

Well Known Business Man Goes To His Last Reward

George W. Hopkins, president of the Union Cypress Company, died yesterday morning at the home of a daughter, Mrs. W. H. Kempfer, with whom he was making his residence here. Mr. Hopkins was born in West Virginia in 1844 and came to Florida many years ago from Michigan. He invested in large tracts of land and timber and founded the Union Cypress Compay of Hopkins one of the largest concerns of its kind in the state. The deceased leaves a widow, a son A. L. Hopkins of Bear Lake, Michigan and two daughters, Mrs. Clare Burgess and Mrs. Agnes Kempfer, both of Melbourne. A daughter died in Big Rapids, Mich. two years ago.

He founded the town of Hopkins which was named after him.

Burial will be in Melbourne cemetery and the funeral will be from the house at three o'clock this afternoon. Rev. Charles Hauck will conduct the services.

George W. Hopkins passed away on January 21, 1925. His obituary ran the next day in the *Melbourne Times*. On January 29, a longer remembrance was published: "Vision, courage, efficiency, persistence, were outstanding qualities in the character of Geo. W. Hopkins. . . . As a boy of ten in the mountains of West, Va., he sometimes stepped out from the rows of corn, and with his hoe laid out in the dust farms of large acreage, with neat houses, barns and fields, apple trees and melon patches. Up among the lakes of Michigan later these dreams came true. . . . In the carefully planned village fronting on (Bear) Lake are neat dwellings and wide streets; a wonderful school house, and where sawmill and lumber yard once were a fine park the pride of the town. . . . On the beautiful ocean prairie west of Melbourne his vision saw well tilled farms, happy homes, contented people. His vigorous efforts to realize this dream were handicapped by Uncle Sam's 'Thou shall not,' but Deer Park and Hopkins have their place on the map, and other towns have their beginning . . . his main business of being a lumberman is another story. Let him be remembered by what he has done." George and Ella Hopkins and their children Clare and Blanch rest in the Melbourne Cemetery next to the Kempfer family. (*Melbourne Times*.)

Schedule A

UNION CYPRESS COMPANY

January 31, 1925.

FIXED ASSETTS;

Pine Mill	$42,713.04
Boarding House Equipment	876.96
Crescent Theater.......	551.69
Cypress Mill(Construction)	106,006.99
Dry Kiln..........	2,727.60
Ditching Yard.............	437.34
Dry Shed..................	1,534.60
Electric Light Plant	2,971.93
Fire Protection	1.129.14
Feed House...................	203.75
Houses.......................	13,298.94
Hopkins Hotel Addition....	559.16
Jane Green Grade............	3,880.89
Live Stock...................	110.80
Log Camp Houses.............	5,502.27
Log Loader.................	1,604.22
Machine Shop.............	3,249.40
Machine Shop Tools..........	1,387.15
New Station................	289.94
Office Furniture and Fixtures	675.29
Pine Equipment..............	12,200.00
Planing Mill.................	13,385.16
Pile Driver................	204.08
Supply House	513.54
Skidder....................	12, 426.42
Saw Mill Derrick...........	4,445.86
Timber.....................	35, 661.12
Track Layer................	3,365.37
Trucks.....................	4,642.65
Wells and Water Works	5,995.85
Yard Construction........	14,034.18
Railroad...................	109,500.63
Rolling Stock..............	30,228.62
Saw Mill	8,701.80
TOTAL FIXED ASSETTS..............	$445,016.38

George W. Hopkins was the controlling factor in building the mill and organizing the company, and mill and railway operations shut down following his death. Total company assets as of January 31, 1925, were $589,934.27. Only the fixed assets are shown here. Claims against the estate amounted to $250,000. Much litigation was still to come. The East Coast Lumber and Supply Company purchased the Union Cypress retail business in March 1925. All company equipment at the mill and on the railway, and all real estate not previously sold in Melbourne, went to his estate on June 1, 1926. All rights to the Hopkins timber also went to his estate, except for that timber on 14,340 acres south of Deer Park and along Wolf and Penna Wa Creeks, which would be sold to the Foshee interests on March 1, 1928. (Katherine Doran.)

The company office and commissary were demolished in 1926 to make room for FEC's new freight terminal. The commissary's grocery business was being moved across the street to the old Crescent Theatre building, which was being remodeled when these advertisements ran in the *Melbourne Journal* on November 14, 1926. C.H. McNulty had bought the complete stock of dry goods, shoes, clothing, and hardware and, on November 17, started reselling it to the public at less than wholesale cost. He also sold merchandise in full case lots to local merchants. The remaining items were then auctioned off daily between December 8 and 24. Thieves looted the store on December 13, taking over $200 in cash and merchandise using tools they stole from FEC's construction site. (*Melbourne Journal.*)

About 37 portable log camp houses, where now located.
approximately 2,800 tons of rail, where now located.
Logging railroad right of way as now located, from the mill site at Hopkins to the land covered by the contract between George W. Hopkins and Union Cypress Company hereinabove mentioned, and said Grantor hereby quit-claims the right of way and any right and title or interest it may have under the deeds conveying said right of way;

Also the following:

1 Double Circular Saw Mill
1 L.H. Band saw mill of steel and concrete construction complete, with all machinery connected therewith
1 Planing mill
2 Dry Kilns
1 Feed House Building
1 Garage Building
1 Water Works system, including 50,000 gal. steel tank, but not including pipe lines connecting up artesian wells and residences.
1 log camp portable commissary car.
6 Locomotives.
45 Log cars.
1 Lidgerwood Skidder.
1 Pile Driver.
1 McGiffert Log Loader
1 Clyde Ground Skidder.
1 Fordson Ground Skidder.
1 Tracklaying Machine.
1 Machine Shop and contents.
1 Supply house and contents.
Stock of merchandise contained in store and warehouse, and any and all other equipment, machinery, structures, or property of any kind now owned by party of the first part located upon the mill site at Hopkins, Florida, consisting of approximately one hundred forty acres, with the exception of the following:

1 Lidgerwood Skidder No. 22996.7.
About 26 houses for whites.
1 store building and warehouse.
1 4-apartment building
1 white boarding house.
1 six room"colored"house.

The Union Cypress filed a voluntary petition for bankruptcy on November 25, 1927. On December 29, 1927, to secure money to pay creditors, the company entered into an agreement with William E. Foshee, president of the Foshee Lumber Company, of Willow, Florida, as trustee, wherein they would sell and he would buy all previous timber "cutting contracts" and most of the buildings and other equipment at the mill site in Hopkins for an initial $58,500. This was contingent upon a new contract for the Union Cypress to sell and Foshee to buy 25 million feet of timber from the Hopkins estate. This "deed and bill of sale" and a companion 14-year timber lease from the executors of the Hopkins estate were signed on March 1, 1928. The six locomotives are unidentified. Glover Machine Works records show that the Union Cypress purchased a used 1907 Glover 2-6-0 from the Port Wentworth Lumber Company in Hiltonia, Georgia. It was shipped in April 1925, but there are no indications it arrived in Hopkins. (Brevard County Miscellaneous Records, Book H, page 186.)

WANTED!

Pine Logs Delivered on Our Tracks

In Our Mill Yard At Melbourne; or log road between St. Johns River and Mill. After Dec. 1st want these logs at Mill Yard only.

Will Pay Following Price; Per Thousand ft. Doyle Scale.

Sound Butt Cut Logs				Sound Logs Other Than Butt Cuts In Same Sizes.
10 to 12" Dia.	14 to 16' Inc.	$12.00	----------------	$10.00
10 to 12" Dia.	18 to 22' Inc.	12.50	----------------	10.50
10 to 12" Dia.	24 to 30' Inc.	13.50	----------------	11.50
13 to 15" Dia.	14 to 16' Inc.	13.00	----------------	11.00
13 to 15" Dia.	18 to 22' Inc.	13.50	----------------	11.50
13 to 15" Dia.	24 to 30' Inc.	14.50	----------------	12.50
16" & up	14 to 16 Inc.	14.00	----------------	12.00
16" & up	18 to 22 Inc.	14.50	----------------	12.50
16" & up	24 to 30 Inc.	15.50	----------------	13.50

Foshee Manufacturing Company

This advertisement ran in the October 5–16, 1928, *Melbourne Times*, while Foshee was in the process of renovating the facilities at Hopkins to complete the mill that the Union Cypress started rebuilding in 1924. New dry kilns, a planing mill, and machine shop were built. The old power plant was demolished and rebuilt to include four 225-horsepower boilers. Thirty-five new houses were built for laborers, and others were renovated. The railway was repaired, and the entire St. Johns trestle was sawed off at the waterline and rebuilt. Operations started January 21, 1929, at the new Foshee Manufacturing Company. By February, over 165 men were employed. Operating expenses were about $1,000 a day, three-quarters of which went to payroll. Standing in front of the FEC-Hopkins freight terminal are, from left to right, Gail Lassiter, railway section foreman Marion Lassiter, Stuart Foshee, mill superintendent Mason Foshee, and secretary and general manager Bruce Foshee. Stuart was a pilot, and Mason held several patents for lumber machinery. (Above, *Melbourne Times*; below, Sallie Madden.)

Locomotive No. 7 is a big mystery. She came to the Union Cypress in the mid-1920s. Her builder, date, and provenance are unknown. She has characteristics of both Baldwin and Rogers locomotives but may be a hybrid, rebuilt by an unknown locomotive rebuilder. An eerily similar locomotive can be seen in the background in several Glover Machine Works photographs, some showing a faded ". . . GAN . . . BER . . . C . . ." on her tender. She became a Foshee locomotive on March 1, 1928. On the end of the train in the photograph above are four old wooden passenger cars that Foshee bought to use as camp cars out in the woods. The seats were removed, and one was converted into a kitchen and the others into bunkhouses. (Both, Sallie Madden.)

Foshee logged long-leaf yellow pine south of Deer Park and along Wolf and Penna Wa Creeks. They acquired six unidentified Union Cypress locomotives in 1928 (see page 113), but apparently used only three. These may have been Nos. 1, 4, and 7. Foshee later sold No. 4 to Peavy-Wilson at Holopaw. In 1928, W.E. Foshee purchased a used 2-6-0 from the Deal-Bachtel Lumber Company in Montgomery, Alabama. She was built in 1907 by the Vulcan Iron Works of Wilkes-Barre, Pennsylvania, for the Coleman Cypress Company at Coleman, Florida. Roughly 60 percent of Foshee's lumber went to South America, Spain, Cuba, Porto Rico (now Puerto Rico), Germany, England, and other countries. About 20 percent went by boat to the Eastern US markets, and 20 percent stayed in Florida. (Above, author's collection; below, Sterling Hawks.)

LOCAL BANK IS CLOSED MONDAY

Merchant's State Bank Closes After Gradual Withdrawal of Funds

The Merchant's State Bank did not open for business Monday morning. A copy of a telegram to the state comptroller, posted on the front door, stated as follows: "On account of continuous withdrawals, unable to open for business Monday morning. Send representative to take charge."

At a meeting of the stock holders

SECOND BANK CLOSES DOORS HERE MONDAY

Melbourne State Has Heavy Withdrawals and Withholding of Deposits

The Melbourne State Bank failed to open its doors Monday morning of this week, making the second bank failure in Melbourne within the past two weeks, the Merchants State Bank having failed to open, Monday, July 15. This leaves Melbourne without

The two stories above were front-page news in the *Melbourne Times* on July 19 (left) and August 2 (right), 1929, but they were not leading stories. Business went on as usual until October 29—Black Tuesday. By July 1930, the lumber market had deteriorated and many companies were cutting prices below the cost of production. Others had already closed for the summer. Foshee had 200 employees and had been producing an average 1.25 million feet each month from January through July. Now operating at a loss, they announced they would be shutting down on July 15. Employees were not required to pay rent for the summer. The July 4 *Melbourne Times* quoted Foshee as saying humorously, he "hoped the light sleepers would enjoy the rest this summer from the whistle." In January 1931, Foshee put 150 back to work five days a week, mainly to provide employment, but the company finally closed in May. A salvage sale was held in December 1932. Every issue of the *Melbourne Times* during February 1933 announced their final sales (below). (All, *Melbourne Times*.)

Beauty may be only skin deep, but rust goes all the way through. Ol' 7, seen above in February 1936, was left to the elements in the weeds at Hopkins. Her ultimate fate is unknown; perhaps she was recycled into new steel during the next world war. Below may be the "4" as an unnumbered locomotive at the Peavy-Wilson Lumber Company in Holopaw. There she suffered this upset sometime in the early 1930s. Recovered and repaired, she continued working until finally put out to pasture at Holopaw. Her final disposition is also unknown. Peavy-Wilson folded around 1954. (Above, author's collection; below, "Frog" Smith.)

Nine

Gone with the Wind

Foshee reportedly had 20 to 22 miles of railway track (40–44 miles of actual rail). He had purchased approximately 35.5 miles of track from the Union Cypress in 1928 (see page 113). When the rail was finally pulled up in early 1933, it was stockpiled at the mill. Bruce Foshee purchased 12.5 miles of that rail from the Foshee company in October to use as collateral for a loan through the Atlantic National Bank of Jacksonville. That rail was to remain at the mill until the debt was fully paid; however, a mortgage satisfaction had not been found.

The Union Cypress railway was neither chartered in Florida nor classed as a "common carrier." The reasons were simple. Logging lines generally were not chartered, because they would then become subject to labor and safety regulations. No charter meant no right of eminent domain. Most lines leased their right-of-way. Purchase was seldom desirable unless the line was going to become a common carrier, since the long, narrow strip of land could be difficult to sell. Being a common carrier also meant having to comply with tough specifications of the state's railroad commission.

Melbourne's Dr. Irving H. Fowle purchased metal roofing, steel trusses, and other materials at the Foshee salvage sale and built a 50-by-100-foot, two-story boathouse on Crane Creek in March 1933. It later served as a boathouse for the Florida Institute of Technology until it was demolished in April 2006.

The Hopkins Post Office remained in service until October 31, 1942, even though the town had been annexed by Melbourne in 1925. Real estate was gradually sold off. During urban renewal in the 1970s, the land was "cleaned up," and the massive concrete footers were simply buried. Some recent buildings have cracked while settling over these buried masses.

Melbourne's future was destined to be written in the stars. On December 19, 1926, the *Melbourne Journal* described studies by Robert H. Goddard in a front-page story headlined "Rocket to Moon Considered a Possibility. Scientists Declare They Will Be Able to Send Shot From Earth To Moon." Why not? Jules Verne did it in 1865 from what is now North Port in Sarasota County!

The boardinghouse (photographed in 1998), the single-family bungalow (photographed in 1999), and eight other residences are the sole survivors of some 69 buildings once owned by the company. The Brevard County property appraiser's records show three houses dating to the early 1910s, six to the 1920s, and one to the early 1930s. The boardinghouse, known locally as the "big house," is the oldest and truest to its original design. Although stuccoed and re-roofed, its interior is virtually unchanged. Three generations of one family lived here until it was badly damaged by hurricanes in 2004. Applications for renovation grants have been unsuccessful. This bungalow and the others are private homes and have been refurbished over the years. (Both, author's photograph.)

These shingles from a 1926-era home could be some of the last ones made before the Union Cypress mill closed in 1925. They are quarter-inch cypress, five inches wide. One is 16 inches long, and the other 13.5 inches. Investigation of buildings damaged by the four 2004 hurricanes also revealed knob-and-tube wiring from Melbourne's first electrical system, installed in 1914, when the Union Cypress was providing the electricity. (Author's collection.)

The Melbourne-Hopkins freight terminal, built in December 1926 on the site of the Union Cypress Company office and commissary, is seen here in January 2004 from almost exactly the same angle as in the photograph on page 33. It had not been used by the FEC for many years and was demolished by the railway in 2006. The property still belongs to the railway. (Tom Byrne.)

When the Macedonia Missionary Baptist Church was built in the early 1970s, the mill's footers were buried and forgotten, only to resurface during the 1999 church expansion. This is part of a large machinery foundation, a mix of portland cement, sand, broken brick, and other rubble, its original shape reminiscent of a large rectangular muffin flipped upside down. The smooth upper two feet was formed aboveground in an open wooden box, 9.5 feet long by 4.3 feet wide, made from six-inch boards. In the ground below that, rough concrete bulged out over a foot beyond the box. That bulge has since broken off. Broken brick indicates recycling, perhaps when Foshee demolished and rebuilt the old power plant in 1928. Holes were left when 11-inch wood piling rotted away. Below, a brickmaker left his fingerprints pressed into the unfired clay of the center brick, their whorls still visible after some 60 years. (Both, author's photograph.)

This prony brake (or de Prony brake) was used to determine horsepower available from a turning shaft or pulley. It consisted of a wood clamp, a 450-pound counterweight (cast in a wood barrel), and a connecting arm. The bracket was bolted around the shaft or pulley, and the weight provided resistance to turning. The amount of deflection was a measure of horsepower. (Author's collection.)

Specially designed refractory brick tiles locked together around the steel I-beams and cooling pipes at the top of Dutch oven furnaces, protecting them from the tremendous heat. Refractory bricks from 25 different manufacturers were found at the mill site. Steel grates with small holes were designed to hold burning sawdust and shavings and still allow air to circulate. Each furnace had 24 of these 36-inch grates. The one shown had been broken in half. These components are shown at the top of page 26 within the Dutch oven below the broiler. (Author's photograph.)

Judge Platt's driveway, off Minton Road, seen here in 1981, used to be the best-preserved section of the old rail grade. His property was the site of Gagner Station (mile 3.4), and his cattle guards contained rail made in the late 1800s. This whole area was bulldozed for a housing tract in 2001. (Author's photograph.)

A legacy of World War I, castor bean plants are southern annuals and tropical perennials. Castor oil and its derivatives have numerous industrial, textile, and medicinal applications, as well as being the bane of whole generations of children. The husk, however, contains about five percent ricin, an extremely virulent poison with no antidote. A potentially lethal dose could be three beans for a child and four to eight beans for an adult. (Author's photograph.)

Probably the best known and most accessible section of the railway was the grade fill and skeletal trestle crossing the St. Johns River and marsh. The eastern half of the trestle, seen at right in 1980, was deemed a navigation hazard and removed in the summer of 2004. The map on page 50 identifies some of the river and marsh features seen in this 1975 aerial photograph. (Both, author's photograph.)

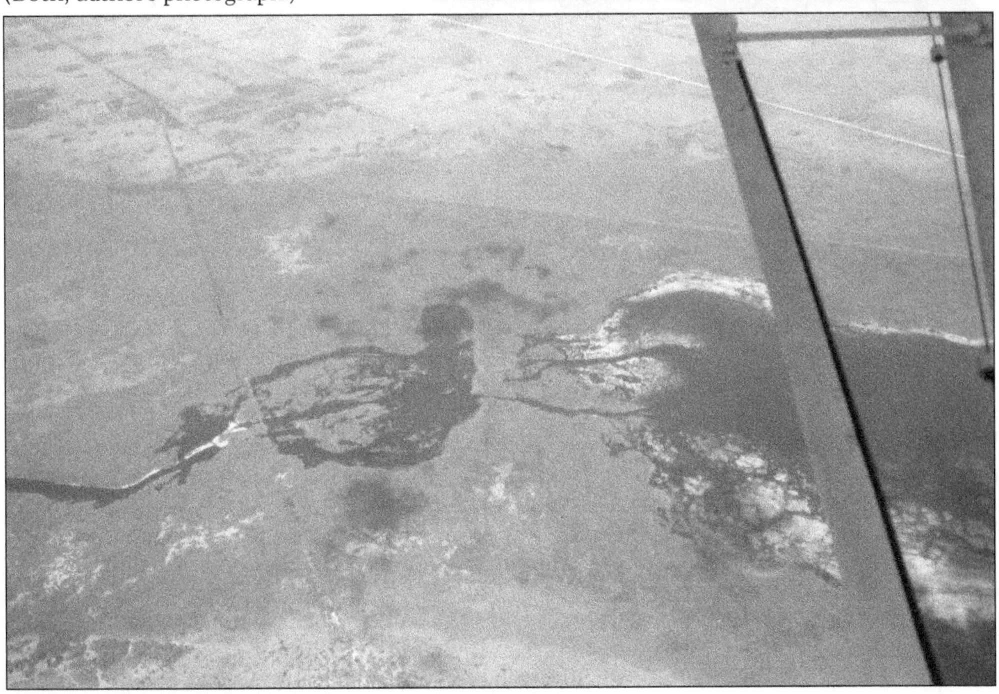

In 1996, after more than 80 years, rotting ties were only about one inch thick at Grassmere Station, on private property west of the St. Johns. The Union Cypress owned its entire right-of-way. Deeds for most of its length stipulated that the land would revert back to the seller should the railway be abandoned. Most of the rail probably ended up as scrap iron for World War II. Those isolated pieces that still exist (see page 46) bear brands indicating they were rolled in years corresponding to the 1883–1886 construction of the BSOC&A. The FEC Railway, University Boulevard, Grant Street, and Line Street define old Hopkins. University Boulevard along the south side of the Union Cypress right-of-way. Florida Institute of Technology's Crawford Building and Skurla Hall sit just north of it. In West Melbourne, Irene Street off Dairy Road, the drainage ditch beside Tallwood Park on Hollywood Boulevard, and Heritage Oaks Boulevard (Gagner Station) and Milwaukee Avenue off Minton Road occupy the right-of-way. In Brandywine Estates, Sugar Creek Drive and Durango Place (Patterson Station) are just north of the right-of-way. (Author's photograph.)

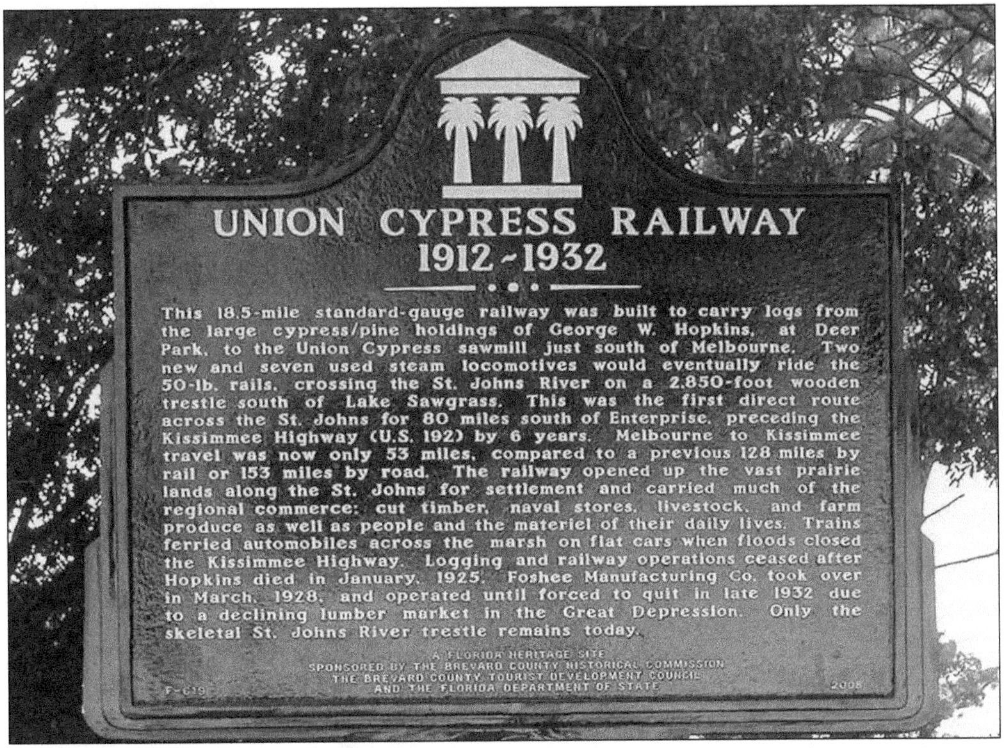

Two Florida state historical markers (written by the author) commemorate the Union Cypress. The railway marker is in West Melbourne's Tallwood Park on Hollywood Boulevard, three-quarters of a mile south of New Haven Avenue (US 192). It is next to the right-of-way, at railway mile 2.8. A large drainage ditch now runs along the center of the right-of-way. This marker was dedicated on October 21, 2008. The sawmill marker is in Melbourne, on Mill Street just east of the intersection of Lipscomb Street. It stands near the southeast corner of the sawmill and power plant. Macedonia Missionary Baptist Church is on the site of the sawmill itself. This marker was dedicated on June 16, 2009. Historical markers are jointly sponsored by the Brevard County Historical Commission, the Brevard County Tourism Development Council, and the Florida Department of State. (Both, Ashley Miller.)

www.ingramcontent.com/pod-product-compliance
Lightning Source LLC
LaVergne TN
LVHW081933200926
848484LV00009B/2416